Gender, Migration and the Dual Career Household

Do careers today demand more spatial mobility? How are two careers accommodated and managed within one household? What issues of gender are related to dual career households and migration?

This book explores the gender issues associated with international migration in dual career households. Adopting a feminist approach, the author links research in economics, sociology, management and business, and human geography to explore post-industrial managerial and professional careers. Particular emphasis is placed on the way in which social mobility and spatial mobility are entwined.

The author explores the location and mobility decisions of dual career households, examining their personal and household biographies as well as published statistics. The couples studied are currently living and working in Canada, the USA and the UK, but have previously lived and worked in many other places.

The chapters examine the following aspects of the subject:

- The micro-sociology of the household
- What it means to be socially mobile
- The complex ways in which work and home are becoming blurred
- Spatial mobility and the acquisition of cultural capital
- Skilled international migration in the pursuit of organisational careers
- Occupational careers and skilled international migration (with case studies of health care professionals).

Of essential interest to scholars of human geography, sociology and gender studies, this book will also interest those working in organisational migration and urban studies.

Irene Hardill's research focuses on the overlapping spheres of labour markets, housing markets and gender relations and has been supported by research grants from ESRC, The Leverhulme Trust, and the French and Canadian governments. In this research she has brought disparate areas of geography together in a study of the socio-economic aspects of labour markets.

Routledge International Studies of Women and Place
Series editors: Janet Henshall Momsen and Janice Monk

Gender, Work and Space
Susan Hanson and Geraldine Pratt

Women and the Israeli Occupation
Edited by Tamar Mayer

Feminism/Postmodernism/Development
Edited by Marianne H. Marchand and Jane L. Parpart

Women of the European Union: The Politics of Work and Daily Life
Edited by Janice Monk and Maria Dolors Garcia-Raomon

Who Will Mind the Baby? Geographies of Childcare and Working Mothers
Edited by Kim England

Feminist Political Ecology: Global Issues and Local Experience
Edited by Dianne Rocheleau, Esther Wangari and Barbara Thomas-Slayter

Women Divided: Gender, Religion and Politics in Northern Ireland
Rosemary Sales

Women's Lifeworlds: Women's Narratives on Shaping their Realities
Edited by Edith Sizoo

Gender, Planning and Human Rights
Edited by Tovi Fenster

Gender, Ethnicity and Place: Women and Identity in Guyana
Linda Peake and D. Alissa Trotz

Gender, Migration and the Dual Career Household

Irene Hardill

London and New York

First published 2002 by Routledge
11 New Fetter Lane, London EC4P 4EE

Simultaneously published in the USA and Canada
by Routledge
29 West 35th Street, New York, NY 10001

Routledge is an imprint of the Taylor & Francis Group

© 2002 Irene Hardill

Typeset in Sabon by GreenGate Publishing Services, Tonbridge, Kent
Printed and bound in Great Britain by St Edmundsbury Press,
Bury St Edmunds, Suffolk

British Library Cataloguing in Publication Data
A catalogue record for this book is available from the British Library

Library of Congress Cataloging in Publication Data
A catalog record for this book has been requested

ISBN 0-415-24173-1

To my parents

Contents

Figures

Plates

Tables

Acknowledgements

This book draws on research funded by the Canadian Faculty Research Program, ESRC (grant R000236072), the Leverhulme Trust (grant F/740), the Ministère de l'Equipement, des Transports et du Logement (grant F00-62), and the Research Enhancement Fund, Nottingham Trent University. The research was undertaken in partnership with Anne E. Green, David W. Owen, Anna C. Dudleston, Stephen Munn, David T. Graham and Parvati Raghuram. I am indebted to the editors of this series Janet Momsen and Janice Monk and to Francine Dansereau, Suzanne Davies-Withers, Isabel Dyck, Kim England, Briavel Holcomb, Eleonore Kofman, Sandy MacDonald, Bob North, Margaret North, Damaris Rose, Joost van Loon, Johanna Walters, Rob Watson, Gordon Winder for their help and to the Department of Geography, University of British Columbia for inviting me to spend part of my sabbatical there in 1999. It was during this sabbatical that I began writing this book. I am grateful to the dual career households who participated in the study, and to Olwyn Ince for her research assistance in writing this book. The views expressed here are those of the author alone.

1   Social and spatial mobility in a global economy

Social relations always have a spatial form and a spatial content.
(Massey 1994: 168)

During the 1990s the male partners of dual career households were elected to be President of the United States of America and Prime Ministers of Canada and the United Kingdom of Great Britain and Northern Ireland. Bill Clinton, Joe Clark and Tony Blair all have partners – Hillary Clinton, Maureen McTeer and Cherie Booth – who pursue successful careers within the legal profession. Moreover in spring 2000 the problems of balancing work and home were placed centre stage when Tony Blair and Cherie Booth's fourth child was born. Cherie Booth spoke publicly about the struggle she has in juggling home and work, and this is in a household that has the resources extensively to commodify many of the tasks of social reproduction (Momsen 1999). The three households form part of a growing minority of households, between 10–20 per cent of all couples in Britain, the USA and Canada, in which both partners are pursuing careers (Hakim 2000: 111).

For the purposes of this book, dual career households are defined as those in which both partners (that is, two heterosexual adults living as a couple in a two person or larger household) are in managerial and administrative, professional and associated professional and technical occupations; the types of occupation which tend to place particular demands on the individual and emphasise commitment (Erikson and Goldthorpe 1992: 42). In this book I focus in particular on the location and mobility decisions of dual career households currently working in Canada, the USA and the UK, highlighting in particular the ways in which social mobility (career advancement) and spatial mobility (geographic migration) intertwine for them. A growing number of households in all three countries are striving to achieve upward social mobility through the pursuit of a career (see below), and are members of the much expanded 'new middle class' (Brooks 2000; Savage 1995).

This deep commitment to the labour market can be examined through the nature of the employment relationship. For managers and professionals, who essentially equate with Goldthorpe's (1995) 'service class' / the middle class,[1] their employment is regulated through a service relationship as opposed to a labour contract. In the latter the exchange of wages for effort is very specific and the worker is closely supervised; in contrast the service relationship is more long term and involves a more diffuse exchange. The service relationship can be understood as the means through which an employing organisation seeks to create and sustain 'commitment' (ibid.). These means include a salary and fringe benefits, such as company cars, private health insurance and performance-related pay.

Both partners in dual career households have a deep commitment to the labour market and seek jobs requiring a high degree of commitment, and of an intrinsically demanding character (Rapoport and Rapoport 1976). Moreover those female partners who are mothers have a stronger attachment to the labour market after childbirth than do mothers in other households. But there could well be a conflict between a life centred on an employment career and the demands of continuous, full-time jobs, and a life centred on marriage, childbearing and childrearing, and the demands of family life. There can be a conflict between production and reproduction as a central activity and principal source of personal identity and achievement (Hakim 2000: 4), especially if within the household there is little accommodation of the woman's career, for example by not sharing the tasks of social reproduction (shopping, cleaning, cooking, caring for children, etc.). Clearly, in socio-economic terms, dual career households form a privileged group, better able to compete economically, and to exercise their influence in achieving priorities, than many other population groups. They have been described as the 'optimum survival kit' household (Forrest and Murie 1987) for the late twentieth century (and twenty-first century). As we shall see throughout this book, for dual career households the intertwining of home and work is particularly complicated.

Social mobility describes the movement or opportunities for movement between social groups and the advantages and disadvantages that go with this in terms of income, security, employment, opportunities for advancement, etc. (Aldridge 2001). Sociological studies analyse social mobility in terms of social class rather than income because income is just one dimension of social position. The schema developed by John Goldthorpe (1995) is derived from measuring an occupation in terms of three criteria: market situation (wage, pension, sick pay, benefits); status situation (status of job); and work situation (level of autonomy/control). An individual may enjoy upward social mobility without necessarily enjoying upward income mobility (to the same degree). Social mobility can be downward as well as upward; there are potential winners and losers. This pattern of upward social mobility in all three countries has been made possible by economic and social changes, which have increased employment opportunities in management and in the professions (see p.11).[2]

Skills, expertise and careers

Both partners in dual career households possess skills and expertise and can therefore strive to achieve upward social mobility through the pursuit of a career. The distinction between economic capital, cultural capital, social capital and political capital as forms of skills and expertise was originally made by Bourdieu in the 1970s, but since has been utilised and developed by other sociologists. Economic capital is the sum of the resources and assets that can be used to produce financial gains, including cash in the bank, vocational and educational qualifications, finding its institutional expression in property rights (Bourdieu 1986). In identifying the concept of cultural capital Bourdieu drew on extensive surveys of attitudes, behaviour and occupational status. He showed that cultural capital and social conventions can serve as powerful barriers to mobility and are crucial to understanding how people internalise class distinctions (Butler 2001). Cultural capital exists in various forms with two analytically distinguishable strains, education and 'taste', which confer the capacity to define and legitimise cultural, moral and aesthetic values, standards and styles. Cultural capital therefore consists of the information resources and assets that are socially valued, such as knowledge of art, literature and music or scientific knowledge. Cultural capital as represented by educational qualifications, especially in non-vocational subjects in the arts and humanities, can contribute to, and shape the style and consumption of the home and in leisure activities. It helps to define a family's position within the cultural hierarchy (see Chapters 3 and 4).

Social capital is the sum of the resources, actual and potential, that accrue to a person or group from access to a network of relationships (who you know as distinct from what you know). Social capital can be used to make money (good social contacts can be crucial to the success of business ventures) or to exert power and influence.

Critically, this involves 'transforming contingent relations, such as those of neighbourhood, the workplace, or even kinship, into relationships that are at once necessary and elective, implying durable obligations subjectively felt (feelings of gratitude, respect, friendship, etc.)' (Bourdieu 1986: 249–50). This form of capital is therefore more of a relational phenomenon than a tangible, or easily quantifiable, resource. Bourdieu has demonstrated the importance of understanding strategies for trading economic for cultural capital and vice versa (for example, the *nouveau riche* financial trader buying an exclusive property, such as a country mansion), and the tensions between different parts of the élite (those whose status primarily rests on cultural capital and those whose status is primarily economic) (Bourdieu 1986; Butler 2001). Bourdieu's work has emphasised the importance of institutions such as the family, private schools and universities in passing on cultural capital and thus in determining life paths. The fourth form of capital Bourdieu identified, political capital, is a special form of connections that arise from contacts developed in other contexts. Societies and individuals can accord different weights to the four types of capital. People poor in economic capital can be rich in other forms of capital.

Careers in the new economy

Since the 1980s economic restructuring has impacted on the world of paid work as economic downturns helped pressure the private sector in all three countries into making longer-term adaptations to the competitive pressures of economies increasingly organised on a global scale (Sennett 1998; see pp.5–6). The public sector has also seen substantial changes, most markedly in the UK which had previously embraced the social security values of European welfare states – these values were developed to a much greater degree than in Canada or especially in the USA (Dixon 1999: 71).

As a result, at the start of the twenty-first century, men and women in Canada, the USA and the UK in both the private and public sectors are pursuing careers in a very different working environment from that prevailing previously. Beck (1992) suggests that in this new world of the risk society, social stratification will be increasingly formed on the basis of 'risk-relations'. Risk relations are marked both by a differential exposure to risks and differential abilities to circumvent or manipulate such exposures. Whereas Beck's main thesis focused on ecological risks, it can be argued that similar processes are at work with regard to social risks (Van Loon 2002, in press). Although Beck is right to argue that scarcity may not itself be the primary logic of the distribution of risk, both 'lack' of risk and 'excess' of risk often correlate with access to economic resources, although usually in a mediated fashion. For example, a large number of managers and professionals are 'exposed' to risks of cancer and coronary heart disease due to 'excess' rather than 'lack' (of paid work, through stress, etc, see below), but they also have greater access to the means of manipulating the risks and their consequences through access to economic resources and social networks (Adam and Van Loon 2000). The risks for women could be trade-offs against greater individual freedom from the constraints of traditional family arrangements.

Pursuing a career (striving to achieve upward social mobility) can therefore be seen as both a specific exposure to risk and as a means of manipulating this exposure because of the ways in which the world of paid work is changing (ibid.). Social mobility is not simply a matter of choice, as Sennett (1998) and others have argued, it is forced on vast populations of the industrialised world as a means of flexibilising the labour force and optimising the distribution of human resources. Sennett's (1998) argument that flexibilisation is a force, not an option, seems to be perfectly reflected in managerial and professional labour markets in North America and the UK.

Flexi-time, flexi-place world

Working life has been characterised by more and more movement. Mobility (that is the social nature of movement) and migration are the 'markers of our time' (Said 1994). This increased mobility and connectedness has produced 'landscapes of mobility' (motorway service stations, bus stations, airport terminals), 'non-places', the 'fleeting', the 'temporary', and 'ephemeral' (Auge 1995: 44). As Cresswell (2001: 19) comments, 'not only does the world appear to be more mobile but our ways of knowing the world have become more fluid'. In the 'new economy' of the Anglo-American world the mass timetable of the industrial world, of the '9–5' office world, and of silent Sundays, has given way to a flexi-time, flexi-place world. The temporal pattern of work has changed in its daily, weekly and monthly rhythms, but so has the 'spatiality' of work: for some paid work is undertaken at home, for others there is more cross-border travel, for others more short-term assignments and for others work takes place in cyberspace. Indeed some writers now use the term hypermobility to capture the above changes (Adams 1999; Cohen 1997). These changes have occurred in the context of the different welfare regimes of Canada, the USA and the UK.

A reconfiguration of the spatial mobility and the temporal flexibility expected of managers and professionals is occurring. Examples include the following:

- An apparent expansion in the spatial horizons of managers and professionals for career development often necessitating international mobility.
- The blurring of business travel, short-term business assignments and residential mobility.
- Trends in the re-arrangement of the spatial and temporal linkages between home and work, mediated by technology and telecommunications, with tasks for paid work undertaken in a variety of locations such as while travelling, whilst at home, and networking whilst socialising.
- Residences increasingly chosen for their access to a number of labour markets (near to motorway hub, airports).

In the 'new economy' the execution of paid work involves more and more movements. The lives of managers and professionals and their households are being transformed by new movements and mobilities.

- Commuting flows have become more diffuse in all three countries and in order to capture their complexity it is important to focus on: the daily, weekly and monthly movements undertaken for the execution of tasks for paid work by working-age managers and professionals; other types of movements as they impinge on the journeys outlined above (such as combining shopping and childcare journeys with work journeys).

- There also appears to be a blurring of business travel with commuting, residential mobility and migration and business travel and migration and there is an increase in use of hotels/rented apartments for business assignments (few days, few weeks, month, year or so), sometimes provided by an employer.
- Moreover the managerial and professional mobility outlined above impacts on residential mobility, with 'place' flexibility an increasing feature of life for managers and professionals (Green 1997), as illustrated by 'commuter' couples and 'astronaut' families (see below; Chapters 2 and 6). There is therefore a challenge to the idea of housing as a united space and for some partners weekends involve movements to 'be together'.

The postmodern household

While the world of paid work has changed radically, family arrangements today are also changing – they are diverse, fluid and unresolved, with a broad range of gender and kinship relations, the 'postmodern family' (Stacey 1998: 17). There is now a greater choice of lifestyle: to live alone, with a partner or with other individuals; to stay single or marry; to remain in or terminate relationships and subsequently divorce/marry/cohabit; to forgo/postpone childbearing or to have children within/outside marriage or other consensual unions. Though greater choice exists, living together remains a conjugal norm, and the most common household form remains heterosexual households. Yet such relationships are often not permanent, nor are they composed of a male breadwinner and female homemaker. Relationships involve 'contracts' between partners, whereas family and marriage imply communal interests.

Generally, relationships formed around family and co-habitation/marriage imply a large element of 'common interests' in order to override individual competitiveness and demands for mobility when decisions concerning the allocation of domestic responsibilities have to be made. Occasionally therefore, one or more partners may be expected to sacrifice (give a lower priority to) their own particular career interests in order to invest more resources in the collective project called 'family' (Beck and Beck-Gernsheim 1995: 52; Giddens 1991). But individual competitiveness and demands for (spatial) mobility are contrary to expectations at home in many heterosexual relationships.

Beck and Beck-Gernsheim argue that 'every marriage consists of two marriages, the husband's and the wife's' (ibid. 62). Living together in those households where both partners 'individualise' may still mean an inegalitarian partnership, with the male partner prioritising his career and the female partner 'juggling' work (often with reduced hours/career break) with 'home' (Blossfeld and Drobnic 2001; Hakim 2000; Momsen 1999). Hakim (2000: 147) recorded differences between wives and husbands regarding marital satisfaction in the UK. The happiest men were those with a full-time homemaker partner, while men in role reversal households recorded the highest levels of marital dissatisfaction. For women the happiest group were those working part-time, while those with a full-time job were more likely to express dissatisfaction because of trying to juggle work and home. In addition juggling work and home can result in some households adopting complex living arrangements, such as:

- being 'shift parents' juggling childcare and new working patterns (Daycare Trust 2000);
- transnational astronaut families with parachute children (see Chapter 6; Pe-Pua *et al.* 1996);
- 'living together apart' as 'commuter/weekend couples' (Green *et al.* 1999; Winfield 1985);
- for a growing proportion of couples, living together without making a lifetime commitment.

Anthropologists draw an important distinction between households, the residential units of everyday life, and families, the more ambiguous, symbolic terrain in which kinship is represented (Moore 1988). Families in an anthropological sense are a locus of meanings and relationships (Stacey 1998: 17). The term household is in fact usually associated with a locus of residence, but for commuter couples and astronaut families there are two places of residence. Astronaut families tend to share resources but not all commuter couples do and for commuter couples their periods apart can be longer than their time together, because of the complexity of lives today we need to revisit these terms.

Career ladders

Traditionally the term 'career' implied some long-term progression, a ladder, or linear promotion, within an occupation, or through a series of occupations involving increasing levels of responsibility at each stage (Evetts 2000), and career progression is closely linked with ideas of social mobility. Linear promotion (thereby gaining social mobility) can be achieved through some combination of factors such as length of service, experience, ability and aptitude, and the acquisition of further vocational qualifications (Bailyn 1993). Promotion ladders can be nationally standardised (as in teaching or nursing) or they can be firm or company-specific. Qualifications and promotions are linked in that occupational qualifications bestow competence on practitioners, which is of great significance to the ideology of professionalism. Qualifications also form the pre-requisites and justification for merit-based systems of promotion (Evetts 2000; Sullivan 1995).

Research has shown that having a full-time job, an uninterrupted work-ing life, and being seen as promotable through having the ability and commitment (which often involves working long hours) to appear as a viable long-term prospect, are key factors influencing an individual's career progression. Career links past, present and future through a series of stages, steps or progressions. A career offers a vehicle for the self to 'become' (Grey 1994: 481), as well as the potential for the management of self through steps on a ladder, but are careers offered in equal measure to both the male and female partners in dual career households?

The 'typical' female career trajectory in Canada, the UK and the USA has been non-linear, complex and dynamic, characterised by access to fewer choices/options (spatial and temporal) and fewer material resources in the woman's life than in the man's (Epstein *et al.* 1999). High status, well-paid jobs tend to be organised as full-time, and are therefore generally incompat-ible with wanting to prioritise both home and work. Moreover women in dual career households, even childless women, are more likely to be the 'trailing spouse' with the 'follower'/ secondary career, which is unplanned and erratic (Bruegel 1996; Hardill *et al.* 1997a). Household migration thus ranks next to child rearing as an important dampening influence on the life cycle wage evolution of married women (Mincer 1978: 771). Typically for the trailing spouse household migration is not associated with the economic betterment and career development that it provides to her partner who has the lead career.

The pursuit of a career begins early, through the acquisition of economic, cultural and social capital with the foundations laid through parental choice of school, in the choice of a specific university, and the choice of a specific course of study. Managers and professionals are both oriented to control and mastery, to creating order through the implementation of an abstract decision process. Both create hierarchical relations to achieve this. The application of expert knowledge, whether confined to the office or the per-son, retains a mysterious character. Careerists include managers (bureaucratic careers) and professionals. The debate on bureaucracy has a long pedigree in sociology, taking as its starting point Max Weber's notion of rational-legal authority, in an office hierarchically arranged, co-ordinated and specialised.

The professions and the professional 'ideal'

The 'professions' are an established feature of the labour markets in the three case study countries. In pre-Reformation England, the honorific pro-fessions – the law, medicine and the clergy – were linked to the then-established institutions of church and state. From the eighteenth cen-tury a central core of national institutions developed to regulate the professions. The British model was replicated in Canada. In the USA, state regulation of the professions developed in the nineteenth century.

'Professional' in the USA designates independent status, however, different from business and wage labour, but less tied to authoritative public institutions than in Canada or the UK (Kloppenburg 1989; Sullivan 1995: 30).

Professional practice differs from the bureaucratic in that professional expertise is derived from a formalised training based on science, as well as the creation of a mystique amongst an élite and the dependence and disabling of those who come to the professional in a client capacity. The profession controls knowledge; it creates specialisations, celebrating depth rather than breadth. Professionals act autonomously, referring a difficult case perhaps to a more experienced colleague, but emphatically not bowing to a hierarchy of offices in the manner envisaged by a bureaucratic model. Professionals offer detached 'understanding' and the portrayal of a professional concern, such as an appropriate bedside manner for a doctor, or keeping emotion at a distance.

The professional ideal, like the bureaucratic ideal, was forged in historical processes where the key actors were men and with cultural notions of masculinity (Davies 1996: 669). Glazer and Slater (1987: 14) link the forging of a concept of profession in the USA with a quest for order in a period of rapid social change and with middle-class male anxiety about proving one's self. Today professionals are no longer sole practitioners, but are part of the workforce of large bureaucratic organisations, whether it is in legal practices, large hospitals or multi-national companies. The purpose of many of these organisations is the pursuit of profit, which does not always support service according to high professional standards. Moreover, advances in technology whether in hospitals, schools, universities or offices, have transformed the nature of the work of many professionals. The medical practitioner, for example, is increasingly tied to the modern technology medical centre.

Medicine, the law, and the clergy still enjoy high status, as illustrated by the fact that when Americans are asked to rank the most desirable jobs, consistently they place medical doctors just below Supreme Court justices, the top-ranked. Lawyers, clergy, dentists, college professors and architects always appear in the top twenty ranked jobs (Sullivan 1995: 2). In the case of the traditional professions clear standards affirm professional status, the effects of official licensing, specialised training and the codification of formalised expertise combined with jurisdiction over vital public activities. In Canada and the UK today, however, there is a crisis within some public sector professions, especially medicine, nursing and teaching, a crisis of recruitment and retention, which will be highlighted in Chapter 8.

In addition to the older learned professions, the professions include engineering and management, and an expanding number of 'emerging' professions, such as the information technology (IT) specialist (Sullivan 1995: 27), most of which require a university education. There is more ambiguity in the public mind as to what other types of occupation are really professional. The picture is complicated by the increasing tendency of managers to seek professional status: witness the explosion of the Masters' of Business Administration (MBA) degree. Modern management clearly aspires toward a recognisable professional identity (ibid. 4; Schoenberger 1997: 140).

Finally, Adonis and Pollard (1998) identify a new, highly prosperous, largely private sector employed 'superclass' within the middle class which stands in increasing contrast to the traditional middle-class professions of, for example, teaching and medicine. Micklethwaite and Wooldridge (2000) also point to the 'cosmocrats', a new global élite, who are connected socially, economically and culturally.

A number of writers have examined professional life as a career often represented as the life cycle model (Huberman 1993) (see Figure 1.1, p.11). Super (1957) for example delineates a series of sequences or maxicycles beginning with exploration and stabilisation occurring at the beginning of one's career. Exploration consists of making provisional choices, of investigating contours of the new profession, and of experimenting with one or several roles. One may move on to a phase of stabilisation or commitment, during which one sets about mastering more systematically the various aspects of the job. For some, this may include a specific focus or specialisation, in the interest of acquiring satisfying working arrangements. For others, stabilisation may lead to added responsibility, with attendant increments in prestige and financial rewards. After this stage routes can diverge, for example by diversification and reassessment.

Diversification may include a search for new stimulations, new ideas, new commitments and new challenges. Teachers for example, may diversify their instructional materials, or methods of evaluation. The reassessment stage gives way to a period of uncertainty; a period of self-doubt may vary from a mild case of routine to an existential 'crisis' over the future course of a career-path. For others, a sense of routine emerges gradually from the stabilisation phase, without their passing through an innovative period mid career (between the 15th and 25th year of a career). The serenity stage is less a phase, more a state of mind around the age of 45–55 years of age. It commonly occurs after a phase of uncertainty or crisis, and entails an ability to accept oneself. Conservatism is a tendency, associated with age, towards increased rigidity and dogmatism, towards greater prudence, towards more resistance to innovations, towards a more pronounced nostalgia for the past. Disengagement is a period of gradual withdrawal and interiorisation near the end of the professional career.

The development of a career is therefore a process rather than a successive series of punctual events. For some this process may appear linear, but for others there are stages, regressions, dead-ends and unpredictable changes sparked by new realisations, in short discontinuities. Discontinuities may occur because of external forces such as accidents, political events or economic crises. Researchers, such as Phillips (1982) in a sample drawn from diverse professions, have used this framework. But it assumes a job for life.

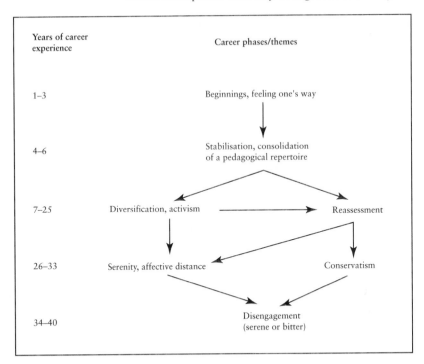

Years of career experience   Career phases/themes

1–3   Beginnings, feeling one's way

4–6   Stabilisation, consolidation
                                 of a pedagogical repertoire

7–25   Diversification, activism ──────────────▶ Reassessment

26–33   Serenity, affective distance   Conservatism

34–40   Disengagement
                                 (serene or bitter)

Figure 1.1 The human lifecycle: a thematic model
Source: Huberman, M. (1993) *The Lives of Teachers*, London: Cassell

Social and spatial mobility

A number of geographers and sociologists have examined the relationship between career advancement (sometimes termed 'social mobility') especially for male workers, within the internal labour markets of large organisations and geographic migration (sometimes termed 'spatial mobility') between different branches of such multi-site organisations (Green 1997; Savage 1988). This work has placed emphasis on the way in which largely male managers and professionals have built a career, achieving social mobility through spatial mobility, typically involving inter-regional moves, with their female partners as 'trailing spouses'. This work makes clear that different organisational and career structures have different implications for the frequency and patterns of spatial mobility. Savage (1988) captured some of these main differences in his characterisation of the three following alternative social mobility strategies.

- Organisational strategy: in which the individual pursues his/her career by moving upwards through the structure of an individual (often large) organisation.
- Entrepreneurial strategy: in which a self-employed individual aims to become a small, and possibly large employer of labour.

- Occupational strategy: in which an individual continually invests in skills-based (often, but not always, occupationally specific) assets – typically gaining experience with a range of different employers – in order to pursue his/her career within their profession.

Each strategy has a different relationship to spatial mobility. Entrepreneurial strategies usually involve spatial immobility, largely because starting a business normally involves drawing on localised resources (ibid.). Organisational and occupational strategies are more complex, but with a relatively high incidence of (inter-regional but increasingly international) spatial mobility (see p.13). And such careers are often built on the foundation of heavy investments in cultural capital obtained within the internationalised higher education system.

In addition, an individual's investment in cultural capital often requires spatial mobility. Therefore for some managers and professionals spatial mobility begins within the higher education system, as indeed it did for the two editors of this series on International Studies of Women and Place, Janet Momsen and Janice Monk. For many managers and professionals their first independent move thus tends to be as a student leaving the parental home and becoming a 'semi-independent adult'. Furthermore, in the last two decades or so undergraduate student spatial mobility has increased with the growth in the number of international student exchange programmes of a limited duration (a semester or so).

The biggest international student movement occurs at postgraduate level, with students moving for postgraduate and professional qualifications, largely taught in English, and some postgraduates combine study with work experience. International migration for postgraduate work is very long established as is illustrated by Ros Pesman's (1996) account of the international migration of Australian women. Flows of graduates from the UK abroad to such countries as Canada, the USA and Australia, or of Commonwealth students to the UK, have also been strong throughout the post-war period.

Career strategies and the new economy

As was noted earlier, managerial and professional careers are being pursued within many restructured labour markets because 'uncertainty ... is woven into the everyday practices of ... capitalism' (Sennett 1998: 31). Managers and professionals, however, are still pursuing what he (ibid. 17) describes as the 'American dream of upward mobility' but not necessarily within the USA. Rather we see:

- work within the internal labour markets of downsized multi-national organisations (re-engineered corporations);

- a more highly qualified workforce meeting different demands: 'commit-ment', 'putting the job first', presentism ('being there') and willingness to be mobile, such as undertaking business travel at short notice, all of which are essential to ensure job retention/career success;
- more self-employment;
- more non-standard employment contracts;
- externalisation of labour and organisational restructuring;
- more labour market transitions, with periods of labour market discon-nection; careers increasingly punctuated with investment in cultural capital through skills updating and skills enhancement;
- the centrality of IT skills;
- more remote information and communication technology (ICT)-based work through the use of laptop computers, fax machines, cellular tele-phones, etc., which 'enable' salaried workers to work at any time, anywhere; including at home, while commuting, or on a business trip (Stanworth 2000: 23; Sennett 1998).

In his recent novel *Super Cannes*, J. G. Ballard (2000) captures some of the key features of the new work spaces. His novel is based on the Science Park Sophia Antipolis in southern France, where 'the corporate city is superbly talented, adult and virtually childless ... [where] work dominates life ... [where] people work harder than they have ever done, and for longer hours. They find their fulfilment through work' (Ballard 2000: 254). Ballard's characters are workaholics, in relationships that are childless.

With the recent shifting of risk in the employment relationship and the extension of instability of employment to the middle classes, the future of the service class has come under scrutiny. Non-standard employment contracts, plus the emphasis on teamworking, initiative, flexibility and adaptability and greater individual responsibility are characteristics often specific to one employer that have changed the employment relationship. In addition, there are also for some new forms of payment – such as that pay is linked to the performance of an individual or to group 'performance targets'. In the new environment the individual becomes an entrepreneur of the self (Rose 1989), indeed Foucault (1979) argues that the self is construed as a self-governing entity. The self makes choices and takes responsibility for action, thought and behaviour, including in the workplace (Giddens 1991: 75). Careers have a particular role to play in such choices, with a focus on the career as an organ-ising and regulative principle – how and why is a career pursued, and how the career is part of the project of the self. The project of self-management links home and work, leisure, dreams and daydreams (Grey 1994: 481) and in this climate of individualisation, agency is more important than social structure (Giddens 1991).

Economic restructuring, especially through the forces of flexibilisation and globalisation, has resulted in other career strategies – in addition to occupational and organisational – including portfolio careers, in which individuals build their own careers, often involving a range of employers/occupations/experience (Ackers 1998). The number of people – both men and women – whose legal and contractual status is one of self-employment has also grown considerably though their actual work situation is far from that of the traditional small business owner (Corden and Eardley 1999: 209).

Entrepreneurial strategies of self-employment are emerging first for managers and professionals (largely male and middle aged) often following downsizing. Increased insecurity in job contracts and dissatisfaction with their terms and conditions of employment are resulting in some managers and professionals moving from employee to self-employed status, often drawing on their social networks and business contacts using their accumulated skills and expertise. Few employ others and they tend to work from home as 'consulting seems the road to independence' (Sennett 1998: 19). Some careerists opt out of the 'rat race' and are attempting to 'take control' of their careers (and workloads) and therefore their lifestyles (Brooks 2000). They trade in high-pressure careers and large incomes for a less frantic and more creative life by becoming self-employed in a different area of activity (ibid.). International migrant professionals form a third group pursuing an entrepreneurial strategy in order to optimise the household's opportunities and advantages (Phizacklea and Ram 1996), often because of labour market discrimination or lack of harmonisation of qualifications in the host country (Ip and Lever-Tracy 1999: 26), especially in the USA (Assar 1999: 96).

Skilled transience and careers

The number of managers and professionals who are/have been/will be skilled transients is increasing. Doyle and Nathan (2001) and Hardill and MacDonald (1998), for example, note a rise in the circulatory movements of British citizens going to live and work abroad for short periods of between one and three years before returning to the UK. Their international migration has been described as 'invisible', because it is perceived not to impose social and economic burdens for the sender and host countries, while often being invisible in terms of ethnicity (Findlay 1996: 515). But occasionally the term 'brain drain' is used when describing the emigration of skilled professionals (see Oommen 1989). The skilled transients move internationally on relatively short-term assignments before returning to their place of origin or transferring to another location, and are now a major feature of global migration systems. But as most skilled migrants are in their twenties and thirties household formation or birth of offspring can impact on their migration trajectories (Ackers 1998; Findlay *et al.* 1994).

Ong, Cheng and Evans (1992) view the ability to migrate as being heavily dependent upon possession of a university degree or similar, since the higher education system has created an international labour market in which occupations demanding high education levels share universal values and a *lingua franca* – the English language. Highly qualified groups such as managers and professionals (possessing a set of skills valued highly by the receiving countries) have been termed a 'club class' of the internationally mobile (Hirst and Thompson 1995: 420) because they are more able than others to take advantage of the opportunities offered by a global labour market. These individuals are said to have formed a new 'international middle class' (see Chapter 3).

Skilled transients move for a wide variety of reasons (Ackers 1998; Adler 1997). Young adults move to combine further study with work experience or for working holidays (Ackers 1998; see Chapter 6). For those employed by multi-national companies – young high fliers and promising middle managers – an international assignment forms an essential part of their career development; the refusal of such an assignment jeopardising their career (Adler 1997; see Chapter 7). For those pursuing occupational careers their professional skills enable transnational job searches but with mixed success for some (including downward occupational mobility, see Chapter 8).

Through a host of linked factors the old style 'expats' whether they be managers or professionals, are being replaced by a number of new figures:

- short-term (brief placements abroad);
- international commuters (regular travel to other offices and sites);
- virtual working (managing a team or project through a combination of travel and remote information and communications technology);
- contractor(short-term projects that require continuous mobility, often filled externally);
- rotator (on/off assignments) (Doyle and Nathan 2001: 6).

For professionals, international migration is influenced by a range of factors, including economic conditions, environmental problems and political change (Li *et al.* 1995: 342), and is channelled by the geography of opportunity, national differences in immigration legislation and the operation of family and social networks (Boyd 1989). Migration is often conceived of as the outcome of a conscious rational decision but it can also be considered as a haphazard, non-rational, non-economic decision. Furthermore skilled international migration occurs within the context of regulatory structures governing migration and career entry both or either of which may present career blocks leading to further migration. For the individual, mobility can be interpreted as an investment in their stock of cultural and economic capital (Bjoras 1989).

Non-economic factors relating to the household as well as the individual careerist can also play an important role in migration decisions. For Bogue (1977), for example, 'social psychological variables' such as the choices, perceptions, feelings and beliefs of the individuals are pivotal in the decision to migrate. His study of migration to Chicago revealed that while economic factors were indeed instrumental, they were not the only ones; family contacts, education benefits, dislike of previous place of residence and a desire to get away from the family all featured as reasons influencing the decision to move. To these could be added individual reasons, especially on the part of a female partner, such as breaking constrictive stereotypes and roles in their own country (Manrique and Manrique 1999: 121). These non-economic reasons can outweigh personal downward occupational mobility and stifled personal aspirations.

Skilled transients are found today in a variety of locations such as advanced capitalist states including global cities; post-Communist Europe; the Middle East, South America and Asia. Thus not all skilled transients live in global cities. As in the colonial era, lifestyles and cultural values of the 'home country' are maintained in an 'expatriate bubble' in some locations, a world within a world, with skilled transients pursuing organisational careers still being mostly male (Adler 1994). It has been suggested that stereotyping on the part of the 'home country' management is a major contributory factor in the exclusion of women managers from international posts with multi-nationals. Women are more visible among those pursuing occupational careers (Ackers 1998; DeLaet 1999; Heitlinger 1999). Skilled transience will be highlighted in Chapters 6–8.

Researching dual career households

In this book I draw on the personal and household biographies of dual career households augmented by published statistics. These couples are currently living and working in Canada, the USA and the UK, but their personal and household biographies have embraced many other locations, for example, a number were born outside the three case study countries and the interviewees' migration trajectories have taken them to most continents. I examine gender as a process in dual career households, something that is active and being accomplished at a variety of sites, including home and work. But I also investigate the impact of social class, in particular the construction and presentation of social class in dual career households (Morgan 1999: 29). Social class has often been subsumed in discussions of race or gender. In this book I seek to address this lacuna by highlighting the life choices of one particular social class, careerist managers and professionals. I also adopt a more fluid understanding of household, as households are processes, not static entities; they are formed, dissolved and reconstituted.

I draw heavily on our original research project, 'The location and mobility decisions of dual career households' (1994–5, funded by the Leverhulme Trust). This work has also been enriched by a second research grant, 'The blurring of boundaries between home and work: intra-familial dynamics and trajectories amongst dual career households' from the Ministère de l'Equipement, des Transports et du Logement, France (2000–1), which enabled me explore the housing implications of the social and spatial mobility practised by British and North American dual career households. My understanding of mobility has also been shaped by my own lived experience, as I have spent several short periods (of about two months) working outside the UK thanks to research fellowships from the French Government (Ministère de la Recherche, 1993 and 1996), and from the Canadian High Commission (2001).

A sabbatical/fellowship abroad is rather like a short-term foreign assignment for someone who works for a multi-national organisation. I, for example, became an 'invisible' skilled migrant, to whom national borders and barriers to the movement of labour did not apply, largely because I am a UK passport holder and am white (see also Wolff 1995). The Canadian Faculty Research programme allowed me to take forward work I had begun in a sabbatical in North America in 1999. Thanks to these two periods in North America I have been able partly to replicate the in-depth interview work I undertook in the UK as well as understand the implications of the pursuit of careers in countries of continental proportions. The interviews in North America were restricted to those working in education and the health sector, and the interviewees were obtained by the 'snowball' method. Their personal stories are used in the book.

With the British study, the in-depth interviews were preceded by a semi-structured self-completion questionnaire survey (for a fuller discussion of the methods see Hardill *et al.* 1997a; Hardill 1998; 1998a). The questionnaire schedule was in three parts, with sections on each career, including jobs held and places of residence since 1980, educational attainment and 'relationship career' (general household information). In most cases the female partner completed the section on the 'relationship career'. An analysis of this longitudinal data set for 130 households is presented in Chapter 2. A sub-set of those who completed the survey element of the research also participated in the in-depth qualitative interviews. Each partner was interviewed separately (Pahl 1989; Safilios-Rothschild 1972).

Most of these face-to-face interviews took place in their homes, and were biographies – life narratives – of a very intense nature (Sizoo 1997). The interview schedules are not a factual record but texts structured by the interviewees' interests, commitments, quirks, etc., 'constructed subjectivity' (McLaughlin 1997: 13). The focus was very much on spatial and temporal dynamics making it possible to identify sources of conflict and negotiation in gender role identities and divisions of labour beyond the 'milestone' events such as marriage and childbirth typically recorded in secondary data and cross sectional surveys. The intention of the face-to-face interviews was not to collect detailed quantitative information, but rather to provide insights into the factors taken into account by dual career households relating to where to live and work. Within migration studies there has been a plea for the use of a biographical approach (Halfacree and Boyle 1993), which I use to facilitate the recognition of the household in terms of a dynamic network of negotiated relations (social, cultural, emotional and economic) exploring issues of mobility and power (Hardill *et al.* 1997a; Jarvis 1997; Jarvis *et al.* 2001).

By using qualitative research methods the inter-related cultural, social, political and economic complexities of dual career households have been mapped (Van Manen 1990; Jarvis 1997). I have thus used lay discourse to construct qualitative and experiential narratives of the different needs, experiences and lifestyles of dual career households (Van Manen 1990). Many feminist researchers have adopted similar smaller scale, qualitative methodological strategies, which aim to break down hierarchical objective ways of knowing (England 1994). Such an approach, utilising more flexible and open methodologies allows one to, 'incorporate difference and acknowledge the partiality and situatedness of our knowledges' (Staeheli and Lawson 1994: 99).

In many instances the accounts of their lives together were told in very different ways by the two partners, and on the homeward journey from the British interviews my co-researcher and I recounted the interviews comparing and contrasting each version of 'their' story. For some of the interviewees the project was therapeutic, and at the end of the interviews we were thanked by them (Bondi 1999). And in some interviews we heard about relationship problems. As a result we found the interviews emotionally draining. A number of the interviews will be presented in subsequent chapters as illustrative case studies of the intertwining of social and spatial mobility and dual career households.

This book however does not focus solely on the household, but also reflects on the human resource implications of careers today. Employers who demand from their managers and professionals 'commitment' as expressed in working more than contracted hours, travel at short notice and assignments abroad, should address the human consequences – on individuals and their households – that flexible capital demands of its skilled workers (Schoenberger 1997; Sennett 1998). Those pursuing organisational careers worked for Japanese, American and European multinationals, with very different policies and practices for skilled transience. And some academics working in the USA had experience of partnership placement programmes.

A number of readers of this book will identify with some of the case studies as they see parallels in their individual biographies, and students may find that their awareness of some of the dilemmas they face, as they try to build careers, will be sharpened. 'Geography matters' (Massey 1985: 51) more than ever today for careerists who are/have been/will be in dual career households.

Organisation of the book

After this introductory chapter the book is divided into eight additional chapters, which explore in greater detail the themes described in this introduction. We begin by looking at social and spatial mobility within the household, exploring the micro-sociology of the household; this is followed by an examination of the ways in which home and work are linked spatially and socially. The third key theme addresses the ways in which cultural and economic capital are accumulated and managed across space.

Chapter 2 consists of an exploration of the micro-sociology of the household, especially relating to careers and decision-making. In Chapters 3 and 4, what it means to be socially mobile is examined, with particular reference to the ways in which partners in dual career households define status and success, in Chapter 3 in relation to careers and in Chapter 4 with reference to the choice of residential property. In Chapter 5 the complex ways in which work and home are becoming blurred is highlighted. The focus of Chapter 6 is a discussion of spatial mobility and the acquisition of cultural capital. Skilled international migration in the pursuit of organisational careers is reviewed in Chapter 7. This is followed by an examination of occupational careers and skilled international migration through case studies of healthcare professionals in Chapter 8. In the final chapter some conclusions are presented along with some reflections on the ways in which employers can foster work-life balance for these 'work rich, time poor' households.

2 Households, careers and decision-making

Introduction

This chapter examines the micro-sociology of the household. It applies neo-Foucauldian concepts of power and negotiation as these are expressed in terms of discursive practices that relate to decision-making about prioritisation of careers. This framework emphasises practices and relations of negotiation, accommodation, contestation, resistance and compromise (Jarvis 1999: 2). Moreover intra-household power relations are also crucial in the understanding of a partners' access to, and experience of leisure, or lack of it (Gilroy 1999: 155) (see Chapter 5). Negotiation is thus explored in very much an interactionist sense, drawing on the notion of human agency (Giddens 1991). Households are seen as dynamic, changing domestic contexts rather than stable, unchanging 'institutions' (Gilroy 1999: 158; see also Ackers 1998; Beck 1992; Beck and Beck-Gernsheim 1995). To comprehend the circumstances, dynamics and trajectories of dual career households it is necessary to discern not only the interaction of housing and labour markets and the configuration of transport systems as well as gender relations, but to understand how dual career households navigate a course through the complex web within which they are 'situated' (as argued by Jarvis *et al.* 2001). In this analysis I draw heavily on the questionnaire survey and interviews with the British and the North American dual career households (see Chapter 1).

In each household structure, a whole host of activities and social relations, beyond those associated with work for economic relations, have to be accommodated and co-ordinated by partners, within particular temporal and spatial constraints. A key spatial constraint is the fixed location of the home, within a particular locale (Hanson and Pratt 1995; Jarvis 1999; Jarvis *et al.* 2001). Household power relations, gender roles and ultimately household structure are deeply embedded within the 'household-locale nexus' (Jarvis 1999).

For career advancement a willingness to be mobile, including to migrate, is often a pre-requisite. Previous studies have emphasised the importance of job-related reasons for long-distance moves, and of such moves being, 'a very positive agent of spiralism for the middle classes' (Johnston *et al.* 1974 cited in Savage 1988: 565). Such terms as 'middle-class cosmopolitan' or 'spiralist' have been coined to describe upwardly mobile, largely male managers and professionals. In the remaining part of this chapter we look at the impact of career strategies of him on her, vice versa and on the household (their lives together).

Theoretical perspectives

There is a growing interest in how individual and household biographies are constructed through time, and Finch argues for, 'an understanding of the social, political and economic contexts in which obligations are negotiated, honoured and abandoned' (Finch 1987: 168). Scholars speculate about the effect that women's formal labour market participation has on the power balance within households, and on the sharing of the tasks of social reproduction (Blossfeld and Drobnic 2001; Gregson and Lowe 1994; Momsen 1999). Phillibeer and Vannoy-Hillier (1990) have shown that there are additional mutual advantages to spouses within dual career couples, which may outweigh the practical inconvenience of having no full-time homemaker, especially when many have no children living at home. Shared social capital, intellectual and work interests can allow spouses greater achievements and upward mobility than they would achieve alone.

Rose and Carrasco (2000) have pointed out a number of ways in which the seemingly equitable division of unpaid labour can be achieved. Negotiations within the household could lead to the male partner's agreeing to take on more domestic tasks. Alternatively, the male partner may be unwilling to take on any additional tasks but the female partner reduces her load by substituting consumable goods and paid domestic services for part of her unpaid household work. Some existing literature, largely British (such as Gregson and Lowe 1995), shows that, in general, wealthier dual earner couples do less household labour overall because they can afford to make such substitutions, and that it is this, rather than the male partner's doing more unpaid work, that is largely responsible for the reduction in the female partner's burden of unpaid housework. In this latter case, there is something of an illusory character to the perception of equity in task sharing within the household.

The 1996 Canadian Census of Population included questions relating to unpaid work, and has thus provided researchers with a unique population dataset with which to explore equity within the home. It was the first national census to include such questions. Analyses of the responses by Rose and Carrasco (2000) shows that while women are still putting in far more hours of housework than men, the differences between males and females are less in the full-time earner categories than in the male full-time, female part-time category. They also found that much less housework is done where there are two full-time workers and no children in the home, perhaps because tasks have been heavily commodified.

Richard Layte (1998: 529) feels that the link between the domestic division of labour and notions of fairness is not derived from relations of economic and quasi-economic exchange, but from the reflexive enactment of gender displays of masculine and feminine accountability. His household analysis (not just of dual career households) used cross-sectional survey data for Great Britain and echoed symbolic exchange theory which views a partners' gender ideologies as more important than economies of labour in defining the perceived equity of outcomes (see Berk 1985). Symbolic exchange theory he felt acts as a residual 'cultural' explanation. However when a woman has 'untraditional' attitudes, these are shown to have a negative effect on her perception of the equity of a situation and make her much less satisfied with assuming a large proportion of domestic work. Thus while more men, on average, are taking on a more active role in the domestic arena than in the past, this is often a 'helping' rather than a 'sharing' role (Crompton 1997; Hardill *et al.* 1997a). Hence, the reality is of many women going home from work for economic relations to a 'second shift' of additional 'work' (work for the tasks of social reproduction). Thus for example, British women in paid work still do three times as much housework as men (Cabinet Office 1998).

This heavy temporal commitment to the home limits the number of hours available per day to travel to work, and to devote to paid work (Hanson and Pratt 1995). For those households with sufficient economic resources some caring tasks can be off-loaded onto someone else, such as a relative or childminder (Wheelock *et al.* 2000). Even if this is the case, women partners still retain the time-consuming and stressful responsibility for arranging childcare, and childcare costs tend to be deducted from their wage (Momsen 1999). Thus for women more than men in dual career households with dependent children the labour market is lived locally (Hanson and Pratt 1995), both spatially (job search area) and temporally (hours per day available to devote to paid work).

This means that many women are left to occupy two roles (Anderson *et al.* 1994): a situation which has been referred to as the 'superwoman syndrome' (see Newell 1993) – or one in which they at least excel outside the home and cope within it (England 1996). Research in the USA (Hochschild 1990) suggests that the increase in the participation of married women in full-time paid employment, in combination with the lack of participation of their partners in the tasks of social reproduction, contributes to the risk of divorce and separation. Gershuny *et al.* (1994; 1994a) has written on processes of 'exit voice and loyalty' over the domestic division of labour, but there is also good qualitative evidence (Hochschild 1990) that complex interactions of attitudes and circumstances can produce varying outcomes in the long run that are not encapsulated in the partner's attitudes.

Decision-making processes

Despite the variability in dual career households as revealed from my British study (Hardill *et al.* 1997a), a number of common factors can be identified in decision-making processes among the surveyed households. When asked to describe the process relating to major decisions one man commented: 'I don't actually feel conscious that there is a sort of process by which we sit down and negotiate'. While one said: 'I am sure in many relationships there is always a little bit more of one person in any decision than the other'. Another said that for their infrequent decisions relating to purchase of cars and houses that 'one of us normally takes the lead'. One man remarked that:

> 'Julia has the view that on all the important decisions, then I usually take the lead and manipulate but she gets away with more on less significant issues. I suppose in the way that I have pursued my career, pretty single-mindedly, dictates the way everything else follows'.

For some households, attitudes to the labour market, in general, and energies devoted to career development and decision-making in the home are separated. For example:

> 'I'm told what to do. I'm under the thumb. But Marie takes all the initiative. I don't want to sound like a wimp, but the fact is that I get all my kicks in the work environment. The conflict and the issues, decisions and judgements and management and everything else. Once I come through this door [home] I have to leave it behind. The kind of job I do is confrontational. Once I'm here it's the family'.

Another man said: 'she probably has a major deciding factor. I'm much more laid back. I tend to go along with things more easily than perhaps she does'. One woman commented: 'we tend to apportion things, I or he will do it'. Another man said: 'we decide everything together, we discuss everything major and as far as I am aware we make our decisions together. We plan out what we want to do'. Another man said: 'we're egalitarian, it's because of our backgrounds and personalities, and because we earn, and always have, within UK £1,000 of each other'. With the decision-making process some partners are 'leaders' and others are 'followers'; while some couples are egalitarian.

When it comes to important, infrequent lifestyle decisions one partner normally takes the lead. Attitudes to achieving in the labour market affect decision-making in the home. Some of the 'lead' career partners also dominated the infrequent lifestyle decision-making process. Interestingly, others with the 'lead' career were happy to adopt a more passive role in other aspects of household decision-making, precisely because they took the 'lead' in decision-making in the labour market (Chapter 4). Power, success and performance in the labour market therefore intersect with 'home' in complex ways.

Career prioritisation

Mincer (1978: 771) has noted that migration ranks next to child rearing as an important dampening influence on the wage evolution of women over the life cycle. Dual career households have to make decisions (either consciously or unconsciously) about whether to pursue both careers equally, whose career should take precedence, and when.[1] Such decisions inevitably involve compromise. Decisions about household migration and location implicitly involve conflict of interests between different household members (Bruegel 1996; Hardill *et al.* 1997a; 1999).

In my British study of dual career households (Hardill *et al.* 1997a) I attempted in the in-depth interviews to designate one partner as 'leader' and one as 'follower' in accordance with whether one career could be identified as leading (i.e. receiving first priority) in household location and mobility decisions. The 30 case study households included 19 instances of 'him leading', five instances of 'her leading' and in six cases no 'lead' career could be identified – in such cases both partners were designated as having careers of equal weighting.

In the words of one male 'leader': 'the job opportunity was king ... everything else was secondary ... it was what I was doing that determined where we live'. Some female 'followers' said they were content, for example: 'I have always put his work first because I can always make mine fit in'. However the majority of the 'followers' were less happy and expressed frustration about the 'sacrifices' they have made. Many of the male 'leaders' were only too well aware of the 'sacrifices' of the followers: one said 'it is fair to say that at every turn she has sacrificed her career'; while another said, 'she has had to compromise and I haven't'. In order to offset the extent of the sacrifice/compromise borne by the 'follower', in some instances the 'leader' was subject to the 'follower's' veto; one 'leader' specifically used the term 'veto', while another said 'I did not take employment in areas of the country perceived as difficult' for his partner's career.

Another way of exploring career prioritisation within dual career households is by the hours per week each partner devotes to paid work. On the basis of hours worked by each individual in relation to his/her partner, again I use the British case study households for illustrative purposes and have divided them into three groups. In the first group, 'male workaholics': the male partner's career was prioritised, with him working over 60 hours per week and undertaking very few of the tasks of social reproduction, while his female partner worked shorter hours and took virtually all the responsibility for the tasks of social reproduction. In these households the women were more likely to work in a part-time capacity in female-dominated professional jobs, such as teaching and nursing, than in the other groups of dual career households identified. It is this group that is characterised by work patterns most akin to those described by Henry and Massey (1995). This group accounted for a quarter of the households in the British study.

The second and third groups represent variations on 'equal partners' model. A degree of distinction is made between the second group, in which both partners invested much time and energy in their respective careers, working considerably longer than their contracted hours. And the third group in which both partners worked only slightly longer than their contractual hours. In the majority of group two households where there were dependent children, some of the tasks of social reproduction were commodified, by hiring a cleaner, child-minder, nanny, etc. (as highlighted by Gregson and Lowe (1994) in their study of dual career households in North East and South East England; see also Momsen 1999). Group two and three households accounted for a half and a quarter, respectively, of those included in the British case study.

Most of the interviewees felt that building a career requires a single-minded determination, Pete[2] for example said, 'I was trying to build a career for myself. Each one was a bigger job'. 'If you put as much into medicine as a career as the men and you are single-minded about it [career development for women follows]'(Suzanne).[3] For James,[4] 'the job opportunity was king, everything else was secondary'. While a career demands 'single-mindedness' the 'lifestyle' they aspired to demanded two salaries (see Chapter 4). But this had not always been the case for some as part-time nurse Helen[5] said, ' we weren't well off then but it didn't bother us the same as it seems to now'. Another interviewee, Pam[6] noted, 'we had tried to manage on one wage, but after I had another child I worked part-time'. Pam's husband Dave said, 'I was the main breadwinner but Pam worked part-time nights for years [to supplement household income]'. In a similar vein Pete[7] acknowledged, 'I don't know how we would have managed [financially] if she hadn't been [a nurse]. Although it is a right bind … it's only because of [her work and salary] that we have been able to [afford to] move [house]'. 'His wages have always been the living wage, we didn't want to stretch ourselves, mine would come in for extras' (Helen, Pete's partner).

The results of the questionnaire survey of 130 households in the British study also shed light on the financial impact of career prioritisation and the lived reality of trying to juggle a demanding job, a meaningful home life and some leisure time. The average hours worked for men in the surveyed households was 46.5 hours per week, and for women it was 36.9 hours (Hardill and Watson 2001; 2002). Table 2.1 presents some descriptive statistics, the means and standard deviations (in parentheses) relating to the pay and hours of work of the male and female partners in all 130 surveyed British dual career households. The males are on average two years older and have almost two years longer job tenure than their female partners and are paid on average almost £9,000 per annum more. On average, the males claim to work almost ten hours per week longer than the females in the sample.

We also provide comparative statistics relating to the 47 households that have never included children, the 33 currently still containing one or more dependent children and the remaining 50 households that did once include,

Table 2.1 Descriptive statistics relating to 130 households

Variable	Males		Females		Household average	
Age (years; *n* = 130)	42.2	(8.5)	40.2	(8.2)	41.2	(8.4)
Age (no children; *n* = 47)	37.0	(8.0)	35.0	(7.2)	36.0	(7.6)
Age (dependent children; *n* = 33)	42.2	(7.4)	41.8	(6.9)	42.0	(7.1)
Age (empty nesters; *n* = 50)	47.0	(6.7)	44.0	(7.3)	45.5	(7.2)
Pay (£000s; *n* = 130)	23.5	(10.4)	14.7	(9.1)	19.1	(10.7)
Pay (no children; *n* = 47)	19.9	(8.7)	16.5	(7.8)	18.2	(8.4)
Pay (dependent children; *n* = 33)	25.9	(9.3)	13.7	(9.7)	19.8	(11.4)
Pay (empty nesters; *n* = 50)	25.6	(9.5)	13.8	(9.8)	19.7	(11.3)
Hours (per week; *n* = 130)	46.5	(10.4)	36.9	(11.8)	41.7	(12.1)
Hours (no children; *n* = 47)	45.6	(9.4)	41.0	(9.1)	43.3	(9.5)
Hours (dependent children; *n* = 33)	46.9	(11.3)	35.4	(13.3)	41.2	(13.6)
Hours (empty nesters; *n* = 50)	47.1	(10.9)	34.1	(12.2)	40.6	(13.2)
Tenure (years; *n* = 130)	6.6	(6.9)	4.7	(5.3)	5.6	(6.2)
Tenure (no children; *n* = 47)	4.6	(6.0)	3.0	(4.5)	3.8	(5.3)
Tenure (dependent children; *n* = 33)	5.6	(5.6)	5.5	(6.1)	5.6	(5.8)
Tenure (empty nesters; *n* = 50)	9.3	(7.7)	5.6	(5.2)	7.4	(6.8)
Job satisfaction (% satisfied; *n* = 130)	92.0	92.0	92.0			
Job satisfaction (no children; *n* = 47)	89.0	91.0	90.0			
Job satisfaction (dependent children; *n* = 33)	94.0	85.0	89.0			
Job satisfaction (empty nesters; *n* = 50)	92.0	96.0	94.0			
Qualification (% with a degree and/or professional qualification)	86.0	85.2				
HE sector (% employed in higher education sector)	38.2	26.4				
Public sector (% employed in public sector)	13.2	22.8				
Managerial (% in managerial jobs)	30.8	19.8				
Professional (% in professional occupations)	56.6	49.2				

Source: Hardill and Watson (2001)

Note: the above figures represent the mean with the standard deviation shown in parentheses

but do not currently contain, dependent children (the 'empty nesters'). These descriptive statistics are helpful in revealing several interesting differences between household characteristics and the pay and labour force participation differences within households. The differences shown in Table

2.1 between the households with and without experiences of child rearing appear to include that:

- the average pay and working hours of the male and female partners in households with no child-rearing experience are not significantly different;
- large differences in average pay and hours of work between the male and female partners appear in households with children, but on average the female partners still devote over 35 hours per week to paid work, as well as being prime carer for their offspring;
- a significant redistribution of within-household career priorities appears to be indicated by the greatly increased earnings and working hours differentials towards the male partners in households with child-rearing responsibilities. The results for the 50 empty nest households also suggest that this radical difference in the average pay and working hours of partners with child-rearing duties, remains even after the child-rearing duties have long ceased;
- while female partners in empty nest households record higher levels of job satisfaction than their male partners, women in households with dependent children were the least satisfied with their jobs as they juggled work and home, often with little support from their male partner.

Commuter couples: living together apart

Deciding where home should be located has traditionally been the male partner's right. In all three countries, the common law that made the man head of household also gave him the authority to choose the marital abode. The female partner was obliged to follow him wherever he went and live under whatever circumstances he chose, with an emphasis on living together and common residence. But historically there have been many circumstances such as war and immigration under which members of the same family, including male and female partners have lived in separate geographical locations. Some workers such as salespeople, truck drivers, business executives, and offshore oil workers to name four, have always left their female partner behind while they organised their lives around work. The military has also demanded frequent separation of couples. People's living arrangements today are very complex, and can be brought about by the prioritisation of both careers, when the female partner wishes to take a job away from 'home', or wishes to remain in an attractive job in the present location if the spouse is offered an attractive job elsewhere. The commuting lifestyle is characterised by an almost total compartmentalisation of work and home life.

Commuter marriage is a situation in which the couple decides to 'live together 'apart'" (Winfield 1985: 4). They have two homes; there are patterns of rent–rent, own–own, own–rent, rent–own, with town houses, condominiums, suburban single family homes, sharing with parents or family, hotel rooms, apartments. Commuter couples who maintain two residences may not pool resources for the running of the two residences in two cities some distance apart. Separations run a continuum from several days apart each week to a month or so as with cross-country relationships in North America where countries are of continental proportions, or with international relationships. In her survey of commuter marriage Winfield (1985) found that one third of couples had dependent children.

Winfield (1985: 14–18) describes four common circumstances in which couples are totally committed to their intimate relationship as well as being totally committed to their careers live apart.

- Young professionals: both partners are trying to build a career and there is no question of one giving up a career if they can't find positions in the same town.
- The relocatees: when the male partner is offered a job in a distant labour market and the female partner does not wish to give up her job and opts not to relocate, the household splits for the furtherance of both careers.
- The well established: those couples who both have important and perhaps prestigious careers in different locations when they develop an intimate relationship, and they choose to continue their two-city lifestyle.
- The economically motivated: a couple split as one partner takes up a job offer elsewhere for economic reasons so the couple maximises earnings and secure a higher standard of living.

Harriet Gross (1980) in her study of the rewards and strains of commuter marriage drew a distinction between two types of couples: 'adjusting' and 'established', on the basis of length of relationship and presence of children. Adjusting couples are younger and childless, both partners are trying to establish careers and may struggle with ascendancy conflicts. While the partners in 'established' relationships are older and have children, their career clashes are less than those of adjusting couples because they have previously established a solid marital reality. Dual career couples who are commuter couples are forced to re-evaluate their lifestyle, including their sexual needs. They are out of traditional role expectations, and traditional economic dependency – two city marriage produces a new social structure created because of the primacy of paid work.

Francine and Jean[8] have been together but apart for 12 years:

> 'It's been like that since we worked together which is 12 years now. It was even worse in the beginning because I was working in … . [Quebec] and he was living in … [Quebec]. But at the same time the kind of work he had at that time, he was doing a PhD but was also working as a consultant, gave him enough flexibility for us to spend more time together.

He could come for two or three weeks between assignments in Africa and I could also come for long weekends. It wasn't that bad in a sense'.

After three years she got a lectureship in the same city as him, and they now have a daughter aged five years so separations impact on their daughter. Francine has periods when she is a single parent:

'She [daughter] finds it difficult when he leaves. Last summer he was away for eight weeks [it] was difficult for her at the beginning when she had to get used to a situation when he wasn't there and then after maybe ten days or two weeks she got used to this and it went along very well'.

Francine then talked about separations from her perspective:

'Given the experience four weeks is something I can accept, which is fine. More than that becomes much more difficult because you have to get installed in a new situation; it's more difficult when he comes back in a sense too. Right now it is three weeks [he was away at the time of the interview], it's just like a weekend for us. But last time was eight weeks and then you have to think in terms of being a single parent for two months, which is very different'.

Beky and Peter[9] ('well established' commuter couple), 'We met in 1993 [through work] at [a] conference ... but became involved five years ago actually, but decided to establish a life together four years ago. We never moved in together because we've never had jobs in the same place' (Beky). When asked if they had tried to secure jobs in the same town Beky commented:

'We have tried. I guess not perhaps in as concerted a fashion as some people might have tried. We took a fairly cold hard look at our situation when we thought about it, not whether we'd be together but whether we could possibly consolidate. We're fairly senior [academics] and most of the appointments are made at a very junior level. I think it would have been very difficult and we probably would have been much more determined to change the circumstances of our jobs if we had been in a 9 to 5, because we would not have been able to have a relationship. But as it is we have enough flexibility that we do actually end up spending the major part of the year together, not necessarily in the same location. But I would say that 60 per cent or 70 per cent even of our time is spent together, but it involves a lot of travelling and a lot of planning'.

Academic flexibility and the ability to organise diaries have helped them maintain their careers and their relationship. Commuter couples stressed the importance of daily communication using e-mail or telephone. This can make living costs expensive. There is also the cost of maintaining two properties and travel times can eat into the time budget. A number of respondents spoke of high telephone bills and those in North America have to spend significant amounts on aeroplane tickets. A number of respondents admitted that sustaining a telephone relationship is particularly difficult, especially where different time zones made arranging telephone calls problematic. As another North American interviewee part of a 'well established' couple put it:

> 'We talk a lot on the 'phone. Everyday, we try to make a connection everyday, but sometimes it doesn't work because there is a three-hour time difference. It's terrible, three hours. It is easier [when I am in Europe for work] it is nine hours [time difference]. So we talk in the morning and the evening is fine. Three hours is terrible. I'm just about to fall asleep, three hours is disruptive. We talk on the 'phone a lot. We don't use e-mail. I cannot use e-mail; e-mail is for business and for relationships with some of my close friends. I can't use e-mail in an intimate relationship' (Hela).[10]

Francine and Jean do use e-mail because his work takes him to Africa where it is not always possible to have access to telephone/cell phones, and so Francine and Jean communicate by e-mail and Jean also sends e-mails to their daughter at school:

> 'Our daughter [aged five years] is too young now to really take part in that, although her teachers in kindergarten have an e-mail address and he sends messages to her, to the teacher. [He] even sends photographs. He has a camera with computer so he can send pictures. For us we keep in touch with e-mail [but this is] not always possible. There are places like last time in villages where it is not possible at all. There was a period of two weeks when we couldn't even get in touch at all. Which is not really something that we like but there was no other way. If something happened he could have phoned I guess, but there is no contact for periods'.

A 'young professional' couple in Canada, with the female partner (Kerry)[11] based in British Columbia and her partner (Dan) a three hour flight away in central Canada, managed to meet up every other weekend, despite the financial cost. They had tried longer separations but that resulted in some relationship strain, so they now meet up more frequently. For others, vacation time was important for longer periods of being together. That is easier for dual career households in countries such as the UK where distances separating couples are less than in the USA or Canada.

Some live together apart because of a job relocation as Simon and Samantha[12] did. The first spell as a commuter couple lasted 18 months when one partner was relocated to a plant several hundred miles away.

They were working for the same organisation, 'we didn't have very much choice in the matter. It is always difficult on a Sunday night to tear yourself away' (Samantha). They have also had a second spell, 'doing the weekend relationship bit again; we hated the fact that we weren't living together' (Simon). Samantha commented on the second period, 'We knew the only thing permanent was our relationship but despite that he went to London to work as a management consultant and I came to Nottingham' (Green 1997).

Childbirth and careers

In spring 2000 the problems of balancing work and home were placed centre stage when one of the UK's most 'visible' dual career couples – Prime Minister Tony Blair and his lawyer wife Cherie Booth – had their fourth child. Both Cherie Booth and Tony Blair discussed in public the merits and demerits of parental leave. Cherie Booth withdrew from the labour market for several months of maternity leave (longer than the statutory period). With the impending birth of a child a woman has to instruct her employer about maternity leave (in the UK) but she and her partner have to reassess home and work. Childbirth is one of the most physically and emotionally demanding periods of a woman's life, and statutory maternity rights and paternity entitlement vary from country to country (see Table 2.2, p.32; Drew 1998).

Childbirth greatly increases the unpaid caring work associated with 'home' as my in-depth interviews with 30 British households demonstrates. Nineteen of these households had children, and in 14 the female partner took either a career break or adjusted her working hours by working part-time. Some of the older women (in their fifties at the time of the interview) did indicate that when they were pregnant their employer had no maternity scheme, one was just expected to 'disappear'. Sarah[13] for example resigned her post when she was pregnant, 'you have to think back to 1969 to 1970 [when] the whole attitude to women with children working was very different. When I think back, women who graduated with me, quite a few of them got married and went [straight] into part-time jobs and never had full-time jobs. That was the social attitude you had in the [19]60s'.

Only one male partner had taken a career break, and another was planning to (see p.34). A number of other male partners took a few days of their holiday entitlement around the time of their child's birth, but most made no adjustment to their working pattern at the time of the birth and afterwards. Sarah said, 'James still pursues his career as if we had no children, indeed as if he was still single, all the give has been on my part'. Paternity rights tend to be viewed negatively in the UK as has been illustrated by the reaction of the Royal Opera House to self-employed opera singer Bryn Terfel's announcement that he planned to take four months paternity leave when his third child is born (Reynolds 2000). While Tony Blair did not officially take paternity leave when his fourth child was born his lawyer wife has

Table 2.2   A comparison of maternity/paternity rights in the UK, Canada and the USA, 2001

UK	Maternity leave entitlement 18 weeks, not linked to length of service with employer. Receipt of Statutory Maternity Pay (SMP) payable for these 18 weeks is dependent on working for employer for 26 weeks plus 15 weeks prior to expected delivery date and paying National Insurance contributions. Additional leave of up to 40 weeks with the right to return to work may also be taken but requires one year's continuous employment to qualify and applies only to mother.
	Maternity allowance for those employees not qualifying for SMP or who are self-employed.
	Maternity grant for those in receipt of state benefits.
	Parental leave may be taken, unpaid, by either parent but they must have one years continuous service with their employer, be the natural or adoptive parent, and give 21 days' notice to their employer. The child must be under 5 years of age. Up to 4 weeks may be taken in a year in periods of one week up to a total of 13 weeks over the 5 year period for each child.
	Parents of disabled children may take time off in periods of one day if needed until the child is 18 years old.
	Emergency leave may be granted if there is a sudden breakdown in care arrangements.
Canada	Maternity leave entitlement of 15 weeks and eligibility is dependent on having performed 600 hours of insured work in the last 52 weeks. Parental leave up to 35 weeks may be taken and if leave is to be shared between partners then each will require 600 hours of insured work. Parental benefits are only payable within 52 weeks following the child's birth or adoption.
USA	No maternity leave entitlement. Time off for childbirth comprises sick leave, annual leave and leave without pay. Women returning to work may claim an advance on annual leave or sick leave dependent on the terms of their employment. Leave without pay, up to 12 weeks, may be granted if all annual leave has been exhausted and must be taken within 1 year of the child's birth or adoption. This applies to either parent. In addition an extra 13 days may be taken by either parent for attending baby clinics or care of the newborn child.
	Sick leave totalling 12 weeks may be granted to either parent if the child has a serious health condition.

Source: UK; www.mymumworks.com Canada; Human Resources Development Canada http: //www.hrdc-drhc.gc.ca/ USA; US Department of Labor http: //www.dol.gov/elaws/fmla.htm

been very vocal in pressing for paternity leave entitlement, which was an election pledge in 2001.

Some women were employed on fixed term contracts when they were pregnant like university researcher Diana,[14] 'a year into my three year contract I became pregnant. I know that [the Head of Department] was most displeased. When he found out that I was entitled to maternity leave I think that upset him even more. He took the view that women are either a mother or a professional woman and you can't be both. He just felt that things were likely to become complicated where a woman had to juggle the responsibilities of a profession as well as looking after young children. He felt that the Department would be short-changed'.

She went on to say, 'whilst I was pregnant I was doing more [work] than I should have done. I was taken into hospital for an emergency C-section because I developed hypertension. I'm sure it was all work-related, [due to] the stress that was on me really. I think words fail me when it comes to describing the pressures that were placed on me, not necessarily verbal. I suppose, like many women do who are pregnant or otherwise, we tend to over compensate and really work flat out. I suppose I was working flat out at a time in my life I shouldn't have been'. Diana has since moved to another university and has secured a tenured lectureship. Although women now have far greater control over their own reproductive body than in the past, Sarah discussed miscarriages, 'I didn't apply for maternity leave when I had my son because the previous pregnancy had miscarried'. Health worries therefore prompted Sarah to resign her post and leave work before she was entitled to maternity leave to try and avoid another miscarriage.

One interviewee, Becky[15] was on maternity leave when she was interviewed, and while she intended to return to work she did express concern as to how she would cope. Prior to their child's arrival Becky had done most of the unpaid work in the home, but Gary also helped, 'a lot'. On maternity leave she had slipped into a routine of having sole responsibility for childcare, acknowledging that she was making 'a rod for my own back'. Like a number of other women interviewed, Becky commented that it was during a period of maternity leave that she started doing more and more household tasks and assumed responsibility for most childcare duties. When she does go back to work she thought it would be hard to break the pattern and share the household work more equitably. Moreover her job as a software consultant involves business trips away from home and she didn't think she would be able to do this. While Gary is a supportive partner there was no way he would care for their son if Becky was away overnight on a business trip.

For some interviewees committing to children through a career break/working part-time for a spell was a conscious choice to enjoy motherhood and they emphasised that they could not see the point in having a child and then not 'being there' for them. Responsibility to 'family' was stressed by Sarah, 'it's a family and you have a responsibility to keep a home and your children need a base'. Thus for Sarah it was important that she invested time in the 'home'. James commented that now their children have left home, 'Sarah's job is probably more important now than it's ever been in our relationship'. Sarah is in an 'empty nest' household, committed to her job and deriving a great deal of personal satisfaction.

Some women felt by prioritising home – because of the gender division of labour in their household – they had suppressed personal aspirations. Claire[16] for example, withdrew from the labour market for five years despite the fact that she had the largest salary, but in their 'traditional' household, Claire was prime carer, with little help from Ian. Once the children were old enough she started working part-time and now works full-time but on a fixed-term contract. Ian is keen they have a third child but Claire is not so keen as it would necessitate her giving up paid work again. They have had bad experiences with child-minders, and this has been a cause of tension in the relationship. The tension is being exacerbated by what Claire sees as Ian's slow recognition of the responsibilities of being a parent although she does accept that he is beginning, 'to grow into fatherhood'.

Only in a minority of households did both partners make career 'adjustments' for childcare. Both Nigel and Anne[17] have taken time out to care for their daughter. Anne was prime carer for one year, 'When I was pregnant with Becky, it was Nigel's first child. He hated his job [social work] and I wanted to get back to work and it was an obvious solution. He really wanted to look after Becky'. Nigel said, 'we actually role reversed as they call it ... for me it was my first child, I felt that by going out to work I was missing out on something important'. He went on to say that, 'we don't particularly want both of us to be working full-time ... the pressures attendant on two full-time careers would be enormous'. Both partners in another household[18] are planning career adjustments for a child they hope to have. Samantha indicated that, 'we will probably try and have children next year. I will take maternity leave ... at the end of those six months Simon will apply for a career break [5 years]. He can't wait – seriously – he cannot wait to give up work'. Simon is feeling disillusioned with his current job.

Cherie Booth regularly talks about the struggle she has in juggling home and work[19] (Bunting 2000). For some interviewees the reality that faces them when they return to paid work after a period of maternity leave is a 'double shift' that can be crippling because of working practices. The realities of uncertain working hours or the need to travel for business including nights away from the family home, result in women committing less to work (and personal aspirations) in an attempt to create a meaningful home life.

Most of the households have commodified some of the tasks of 'home' (cleaning etc.), and use is also made of labour saving devices. But the lack of quality of their home life following childbirth is the reason why some women began to work part-time, as the example of Jane[20] illustrates. Although she and Mark are an egalitarian household when it comes to decision-making, they both work unpredictable hours. While Jane works from a fixed base, Mark travels a lot within the country. When she first returned to work after maternity leave it was to her full-time post, but as Mark's hours were more unpredictable than hers, she was responsible for taking their child to nursery and collecting her in the evening. At work she used to get anxious that she would not be able leave with sufficient time to travel to pick her daughter up before the nursery closed. As a food technologist she worked on a production line and could never exactly predict when her working day would end.

The pressures of this uncertainty resulted in her switching to part-time (21 hours a week), with hours she feels are 'about right'. She did comment that her employer is 'prejudiced against part-time work'. Thus childbirth can impact negatively on career development often because employers and/or colleagues interpret a career break/part-time work as committing to home rather than to work (Epstein *et al.* 1999).

Joanne[21] as a General Practitioner also talked about her decision to work part-time after the birth of her son. 'The original agreement of the contract was that I would do half time until the child or children went to school and at that point I would have the option of going back full-time if wanted to. The average working week is 80 hours, I was not prepared to do that if I had a family, I didn't see the point in having children and leaving them with somebody else. I do 40 hours with the practice, the rest of the time I'm at home or doing my own thing, or doing paperwork for the practice, all this administrative stuff ... with nights and weekends it is 60 hours'.[22] Thus even though she is part-time the number of hours she devotes to the practice is 60 hours per week, considerably more hours per week than a typical full-time contract of 37 hours. Joanne created space for her family by working part-time.

Conclusion

From the evidence collected in in-depth interviews with partners in dual career households, although both partners attach importance to, and devote much energy to, pursuing their respective careers, in most households one career – generally the male career – tends to be prioritised. These are the very households that appear to 'individualise' the most as they adopt more 'traditional' gender roles (see also Dormsch and Ladwig 1998), and evidence from Canada and the USA reveals a broadly similar situation (Rose and Carrasco 2000). In a minority of the surveyed households both careers had equal weighting (especially those with partners under 40 years of age) but this can mean living together apart.

'Commitment' as expressed in holding a full-time job, working above contracted hours and taking no career breaks results in higher earnings. Working part-time or taking a career break though done for short-term convenience (such as wanting to prioritise home and be there for the children, or trying to juggle work and home with little help from their partner, or moving to another part of the country/abroad because of a partner's new job) is likely significantly to affect a couple's long-term financial health, pension entitlements, etc.

The career histories described in this chapter illustrate that the expanding temporal and spatial dimensions of the working environment (unpredictable hours, business travel) in households where both careers are prioritised can mean a commuter marriage. In households with children, even in what had been egalitarian relationships, it is women rather than men who adjust their working hours, and consequently put their careers 'on hold'. While employment policy in all three case study countries has certainly begun to tackle patriarchy in the workplace, the new working environment of commitment to giving more and more time to the job means that the careers of men and women, especially those with children, are following different trajectories. The reality of 'home' is somewhat different to what Anthony Giddens (1998: 93–4) advocates in the Third Way. He writes about democratising family life, which might 'combine individual choice and social solidarity' embracing, 'shared responsibility for childcare'.

3 Defining status and success through the pursuit of a career

Introduction

In the preceding chapter the difficulty in prioritising both careers to the same extent was examined, in this chapter it is probably the peculiar British obsession with class (Cannadine 1998) that has made me wish to reflect more on what the pursuit of a career means for partners in dual career households. Rather than critique approaches to class, in this chapter I highlight what it is to be a manager or professional in terms of status and success, with particular reference to the pursuit of a career and in the following chapter with the purchase of residential property. The traditional Marxian approach to class is based upon the labour process, and how workers relate to capital (and the bourgeoisie). Managers and professionals arguably are already outside the 'working class' as their careers place them as members/potential members of the bourgeoisie, so they are bound up in efforts to step out of the category of worker, and into the category of capitalist. Thus for careerists there is a 'middle-class' angst of working/ career/status, as well as an interaction with ideals of consumption (Butler 1995: 3).

In the Weberian tradition the type of work an individual performs, or occupation, defines the individual's location within an industrialised society's social class system, indeed some argue that 'occupation provides the most powerful signal indicator of social class' (see Johnson and Bowman 1997: 201). Some occupations demand greater skill, longer training and higher levels of education than others do. These factors are significant in determining the social prestige given to an occupation, and are connected with income, place of residence, mode of life and social standing, in sum 'his [sic] socio-economic class position in terms of the whole society' (Susser and Watson 1975: 112). Occupations requiring greater skill, longer training, and higher education than others receive greater rewards and prestige.

Current discussions of class include the 'economy accounts' of neo-Marxist writers discussing post-Fordism (see Amin 1994 for an overview) to the more 'cultural' accounts such as the arguments of Giddens (1991) and Beck (1992) concerning the development of 'reflexive modernity'.

There have been attempts to combine Weberian and Marxian conceptions of class, as illustrated by the work of Mike Savage (1995). Feminists have pointed to the limitations of the dominant versions of mainstream class analysis (Marxists, neo-Marxist or Weberian), which define class purely in terms of labour market positions and derive women's class position from that of the male head of household (Delphy 1981). Others argue that not only are social classes gendered but so are the processes of class formation (Crompton 1997). Rosemary Crompton, for example, argues that the service class is a gendered class not just because of its dependence on female white collar workers but in terms too of its reliance on women undertaking the tasks of social reproduction (see Chapter 1; Gregson and Lowe 1995: 149–50; Mattingly 1999).

Sennett (1998: 65), perhaps reflecting the individualism demanded today, also argues that class involves a far more personal estimation of self and circumstance; the American obsession with individualism expresses the need for status in these terms; one wants to be respected for oneself. His recent writings on careers are interesting in that he repositions the word 'career' into a working-class ideal, one that as a consciousness can be carried over into a middle-class career (ibid.). But with this individualism also comes anxiety, feelings of disillusionment and fragmentation (Beck 1992; Sennett 1998).

Moreover both Beck (1992) and Sennett (1998) point out that work pressures and the culture of individualism can impact negatively on personal relationships. Sennett (1998: 26) acknowledges the conflict between family and work, asking 'how can long term purposes be pursued in a short term society … which feeds on experience which drifts in time, from place to place and from job to job'. Beck and Beck-Gernsheim (1995: 62) argue that 'every marriage consists of two marriages, the husband's and the wife's' (for a fuller discussion see Chapter 1). The culture of individualism in the work place has coincided with the failure of relationships for growing numbers of cohabiting and married couples in Canada, the USA and the UK. But it is in households where one partner works in the armed forces that some of the highest divorce rates are recorded. Military careers particularly demand a high degree of spatial mobility with long periods of separation (Hansard, 2 November 1999; Hardill and MacDonald 1998; see Chapter 7).

Finally some authors explore the impact of globalisation on class (Antonopoulou 2000). Globalisation also involves the transformation of the social relations of production on a global scale. Breen and Rottman (1998) argue that a 'new international middle class' is challenging existing notions of class, which have usually used the national state as the geographical unit for class analysis. Discernible in this transformation is the emergence of a privileged and mobile international capitalist class, composed of skilled professionals but also entrepreneurs with capital, who can purchase immigration and citizen status through business immigration categories in countries such as Canada (Harrison 1996: 7). There does seem to be an emerging global capitalist class whose status symbols, interests and lifestyles are broadly similar (Mitchell 1993). The emergence of business immigration programmes (for a fuller discussion see Chapter 6) is symbolic of the emergence of this class, and its cosmopolitan belief that anything can be purchased, including citizenship (Harrison 1996: 18). We now turn to an examination of identity and success through the pursuit of a career.

Identity and work

An individual's identification with work is complex. Why people work and what they expect from work, and how much commitment they are prepared to put into work, varies from person to person. Work is at the centre of many careerists' lives today, and has taken on an unprecedented individual, family and social significance (Reeves 2001: 116). It is where we spend more and more of our time, and it is absorbing us (ibid.). Judith Doyle (2000) argues that there is a growing divide between those who identify with their work and those who do not. She suggests that some people love their work and identify strongly with it – so much so that there is no boundary between work and home. With an increased perception that the work is riskier and more uncertain than ever, some individuals are searching for a place of belonging. Some have located that place in work. Work for them provides appreciation, recognition, some control, some self-expression, and often a real sense of security. These workers may be dubbed 'willing workers', and for them work is good. Conversely, for the 'wage slaves' work holds no appeal. They work because they have to and often their jobs lack all the criteria of a 'good' job.

Whether they are 'willing workers' or 'wage slaves' managers and professionals do exert some autonomy as to how, where and when their work tasks are executed (Chapter 5). The briefcase has become one of the most recognisable emblems of this autonomy (Sullivan 1995: 1). 'Every working day the briefcases and their bearers … emanate an alluring scent of success' (ibid. 1). They are the signs of a certain status announcing that the bearer holds an occupation sufficiently responsible that work (for economic relations) is not confined to a production shift/opening hours. But the bearer commands one of the rarest forms of credit: autonomy in work. The briefcase asserts that this person can be counted on to do the job without direct supervision, out of the workplace, and off-hours. If asked to name this type of highly valued, highly trusted work, most of us would say 'professional work' (ibid. 1–2).

Dual career households express great fear for the future, perhaps because they are the very ones who have a lot to loose with redundancy (Dean 1998).[1] The status and identity of careerists is intimately bound up with their paid work, so if they loøse their job, they loøse their social status and the material means to support their identity markers (house, car, private education of children, etc.). The loss of paid work can cause an identity crisis. This is well illustrated by Gerald in the film *The Full Monty*. Gerald is unable to tell his wife that he has been made redundant, and so goes through the routine every morning of leaving the house dressed for work clutching his briefcase and sandwiches. His briefcase is his marker of status and autonomy, but it is empty. But as the steelworks where he was in a position of responsibility is no more, work does not exist any more, so he spends his day at the Job Club. This image of struggle is also true for Japan, where unemployment is a new phenomenon.

But what it means to be 'professional' is changing. In the 1980s, Louis Auchincloss (1987) in *Diary of a Yuppie* witnessed a new form of professional, 'mercenary professionalism', with a devouring orientation toward career success through instrumental achievement. The ethos of yuppie life is one with material security and opportunities for personal fulfilment, but it is characterised by intense competitive pressure and little free time (ibid. 132). Yuppie life has harsh dichotomies, competence and adaptability, but detachment. The yuppie must 'travel light', emotionally as well as physically.

More recently David Brooks (2000) in, *Bobos[2] in Paradise*, identifies a 'new' American bourgeoisie (bourgeois bohemians) for whom, 'work … is a vocation, a calling … employees start thinking like artists and activists, they actually work harder for the company. In the 1960s most social theorists assumed that as we got richer, we would work less and less. But if work is a form of self-expression … then you never want to stop'. Brooks' argument is that non-economic motives of self expression also shape people's career decisions. The US-based hospital drama series *ER* (based on the UK series *Casualty*) presents images of dedicated, committed professionals, doctors and nurses, for whom work in a public hospital in Chicago is almost a vocation, as does the school drama *Boston Public* for those employed to teach in a 'challenging' school.

Some studies of middle management however have suggested that there is growing disillusionment and disaffection amongst managers with some feeling themselves to be 'victims' rather than the 'architects' of change (Redman *et al.* 1997). There is talk of 'burnout', 'professional suicide', and 'mid-career crisis' (Hunt 1982; Masluch 1987); the 'managerial menopause' (Davies and Deighan 1986). These studies highlight that commitment has its consequences!

Defining status and success through a career

In this section the interview transcripts are examined to explore narratives of expectations, strategies, dreams and denouement relating to careers. Richard Sennett (1998) positions careers with expressions of personal and social identity built over a lifetime, but very few of the interviewees had a clear long-term career strategy as Samantha,[3] a product development manager in the UK did. She began laying the foundations of her career during her final year at university, 'I definitely had a plan, how I was going to choose a career. So I spent more time doing that than I actually did in studying'. Moreover her first post gave her an early career boost, 'because I was on this fast track and it gave me a lot of exposure to different functional areas'. But then she did not progress fast enough, 'I was with them seven years. Basically I just got bored … there was still an attitude that you'd got to serve your time. Me, like a lot of people, I guess in Thatcher's generation, decided we didn't want to wait that long, we want to do it now. If we are capable of doing it why can't we do it now? A lot of us actually looked and moved to other companies'. Samantha's description of her career strategy chimes in with the narrative in the *Diary of a Yuppie* written in 1987 by Louis Auchincloss. She stresses that her attitudes to work were shaped by the culture of the 1980s as she identified herself as one of 'Thatcher's generation'.

She also went on to describe herself as 'quite hard-nosed'. Not only did she have a clear strategy for herself, but also for Simon, her partner, 'I have actually decided now that we have got a big choice. We can either move on upwards – upwards and onwards – now in pharmaceuticals. Good jobs are all in the boardrooms, and we thought should we do that.' When asked if Simon had supported her career, she said, 'I'm not where I am because I've got a supportive partner, but I'm still with Simon because I've got a supportive partner.' In Simon's interview he acknowledged that, 'career is more important to Samantha than it is to me. [She] wants to have a specific career … a progression in her career'.

While some women partners had a clear career strategy, and held the lead career in the household, most often the female partners prioritised his career to the detriment of theirs, as is illustrated by Pete[4] and Helen. Pete said, 'we decided that Helen's career would be on the back burner … and it hasn't affected where we've looked at jobs … where I work [is] more to the point'; while Helen said that he was 'trying to make a career for himself'. Helen refers to Pete's career rather as a personal quest than a strategy for the household. As a result of Pete pursuing a career, he has worked for five different companies, securing jobs with increasing levels of responsibility in four different regions in the UK in their 25 years together. With each inter-regional move they have had periods as a (relocatee) commuter couple. And Helen as the trailing spouse has, 'worked for five different Area Health Authorities … because of his job'.

Although Helen's career has not been prioritised her salary has been vital to the household as Pete acknowledged, 'I don't know how we would have managed [financially] if she hadn't been [a nurse]. Although it is a right bind … its only because of [her work and salary] that we have been able to [afford to] move [house]'. While the moves had been initiated because of his career, the choice of property was also important for Pete and he was keen to climb the property ladder, so now they occupy a 'character' property with land in a small village near to Nottingham, UK. For Helen the last three inter-regional moves were described in relation to their impact on the children, 'traumatic for the children … not popular' rather than on the accumulation of material assets.

Some of the 'trailing spouses' spoke about their frustration with stifled personal aspirations, as did Claire[5] for example. She withdrew from the labour market for five years following the birth of their two children; despite the fact that she earned more than Ian did, his career was prioritised. The loss of Claire's salary meant the need to find 'creative ways of meeting our needs' including car boot sales. Once the children were old enough Claire started working part-time and now works full-time but with a fixed-term contract.

The majority of the careerists interviewed, both managers and professionals, knew clearly what area of work they wanted to be in. Beky[6] for example, at an early age developed a deep interest, a deep love of her subject: 'I think when I was 12 I became very interested in human evolution'. Two years of not working in a university department confirmed to her that she must get back into academia: 'at the end of two years I felt very strongly that I had to get back to academia'. Francine[7] described her work as a demographer, 'I really I think I had a passion for this study'. But she has made job relocations within Quebec for personal reasons, moving from the Francophone world to an Anglophone institution to get a job in the same city as her partner.[8]

Most had no clear strategy or plan for their working lives; they talked of 'drifting' into jobs, of serendipity, of accident. A number when explaining why they held their current post talked of applying for a number of jobs, and of being in their current job because this was the first employer to offer them a post. They felt they couldn't refuse the offer, and hold out for an interview with their preferred employer. Others talked of intervening circumstances, often personal. Beky had secured her ideal first post in a British University, but resigned it because her then partner (an Iranian passport holder) was encountering residency problems in the UK: 'although the post in [university in the UK] was a temporary post, it was only for three years. There were possibilities. I really had come to feel at home in Britain, my son was born there. His (her husband's) immigration status [in the UK] was so tied to mine [we decided to] come back to Canada and see if he could establish himself in Canada. But we did separate not long after coming back to Canada'. Beky resigned her post in the UK, and came back to Canada and an uncertain future professionally for the wellbeing of her partner.

Sennett (1998) in his recent book described the collapse of the 'conventional career' through the downsizing of IBM in up-state New York (see also Arthur *et al.* 1999). A number of the dual career couples surveyed also experienced downsizing at first hand. Indeed a number of interviewees talked about labour market uncertainty, of having a post that required them to make people redundant, but then being on the receiving end themselves. During James's[9] career, for example, as a production director with at least three companies, felt he had 'specialised' in making people redundant but with his last two employers he was also a victim of downsizing: 'that was quite a shock to the system to be on the receiving end'.

The partners in households with non-standard employment contracts certainly did not talk of a career strategy. For them, keeping in work was the important task, and ideally they wished to secure contracts in the same region/town so they did not have to be a commuter couple. Jason and Caroline,[10] for example, have been together for 11 years. They met at university when she was working as a research assistant and he was doing post-graduate work. He has held a series of contracts in private sector research laboratories and in higher education, but also been unemployed twice. Caroline now has a permanent post in Birmingham and Jason currently works in Nottingham, which means a two-hour commute every day on congested motorways. Caroline does not want to keep moving around, and would love to put down roots; while she has found a secure job this is not the case for Jason. His current contract is fixed term, tied to a research grant which ends in 12 months time.

David Brooks (2000: 108–9) has suggested that today motives for self employment are complex, 'business is not about making money; it's about doing something you love. Life should be an extended hobby … in this way business nourishes the whole person'. He suggests that a counter-cultural mental framework has come to the business world (ibid. 111), but with the information age élites, such as those in high tech Silicon Valley there is a 'hybrid culture … which mixes antiestablishment rebelliousness with Republican laissez faire' (ibid. 111–12). A subset of the interviewees was clearly identifiable as those who had decided to 'opt out', and used self employment as a way of creating a meaningful life. Gerry,[11] a freelance editor, became self-employed because his employer, 'relocated the operation back to London. I wasn't going to start commuting[12] again, so I took the voluntary redundancy option and decided to go freelance. I'd been doing a certain amount of freelance anyway. So it was just a question of moving over to full freelance'. He is in a profession with a long tradition of freelance work (Baines 1999). He noted: 'book publishing is in turmoil at the moment, so nobody is investing in staff. The trend is away from in-house staff, more towards outworkers who are engaged just for the period that they are required, … because of the technology we no longer produce proofs in the old fashioned way … it's all done on screen now'.

Entrepreneurs are an immigration priority category in countries such as Canada but not currently in the UK[13] (Harrison 1996). Their skills allow them to 'make money' in another country. Canada implemented a Business Immigration Program in 1978, initially in order to address a growing fiscal crisis. The wealth they brought was viewed as one means of stimulating economic growth and creating jobs (ibid. 19). Canada's policy is similar to that in the USA and Australia, and reflects more general trends in immigration policy under globalisation. Business immigrants (self employed and entrepreneurs) still qualify under the point system, but their business skills are a primary entry criteria. A third category, 'investors' (having a net worth of at least half a million Canadian dollars) was added in Canada in 1986.

The annual numbers entering Canada as business immigrants amounted to 6,225 (7 per cent of all immigrants) in 1983 but by 1992 had risen to 28,143 (11.1 per cent of all immigrants) (ibid. 12). Between 1983 and 1992 36 per cent of all business immigrants to Canada (40 per cent of all entrepreneurs and 48 per cent of all investors) came from Hong Kong. Their migration was undertaken because of worries regarding the handover of Hong Kong to China in 1997 (ibid. 17). Other countries (Australia and the USA) also made amendments to their immigration rules in order to capture Hong Kong's threatened wealth. While some argue that such programmes commodify citizenship, 'being Canadian has no more cachet than possessing an American Express card' (ibid. 19).

Globalisation does appear to be strengthening a unified international bourgeois class but other migrants adopt an entrepreneurial strategy upon arrival in a host country, and do so for a number of reasons. One oft-cited reason is as a 'survival' strategy to optimise the family's opportunities and advantages because of labour market discrimination (Phizacklea and Ram 1996). Another is the lack of harmonisation of qualifications (Ip and Lever-Tracy 1999), especially in the USA (Assar 1999: 96). But entrepreneurship can be the 'choice' of well educated second generation ethnic minority members, as illustrated by those exploiting diasporic business connections (Hardill and Raghuram 1998). A recent study of Asian (Indian and Chinese) migrants to Australia revealed that their high level of educational attainment and self employment was being used as a strategy to optimise their family's opportunities and advantages (Ip and Lever-Tracy 1999: 76). It cited the case of a medical dual career couple, both with unrecognised medical qualifications from China, who had opened a joint practice where he provided acupuncture and she prescribed and sold herbal medicines (ibid. 68).

A second example is that of the US motel business which is commonly referred to as the 'Patel-Shah' industry, as Indian migrants – predominately Gujaratis – control 12,500 hotels and motels (Assar 1999: 82). These hotels are family businesses, using family labour and community resources (ibid. 99). Some of these Gujarati migrants 'first immigrated to Canada where immigration laws are less stringent. Then they simply crossed the border as Canadian citizens to do business in the USA without restriction' (ibid. 96).

Conclusion

This chapter has explored what it is to be a manager and a professional in terms of status and success, with particular reference to the pursuit of a career. The social status of managers and professionals is intimately tied to their position in the labour market. A diversity of career strategies have been identified, but for only a minority of the interviewed dual career households could their career have been described as a 'well made road'. Indeed very few of the interviewees interviewed had a clear career strategy. Drift indeed perhaps more accurately describes their careers, the result of a combination of factors including the precarious nature of employment, juggling two careers within their household, as well as the interplay of housing markets and labour markets, with personal mobility being shaped to some extent by the property market. While some careerists 'live to work' others 'work to live', and still others 'work to survive'.

4 Defining status and success through residential property

Introduction

Middle-class attitudes to housing have changed in a number of respects. First, with the Industrial Revolution there was a shift of paid employment away from the house. Second, with the increased separation between home and work, the home has taken on new meanings for both middle-class children and adults (see below). It has become a haven of family life protected from the stimulation and threat of the city (Harris and Pratt 1993: 281). Third, it has become an important status symbol, a measure of personal success. In this chapter the multiple meanings of home are explored for the middle classes in general, and for dual career households in particular. Middle-class and dual career households are distinctive in that they not only regard housing as an investment, but also imbue it with cultural significance. Marxists view home ownership (and bourgeois–led suburbanisation) as a solution to certain 'contradictions' of capitalism, the necessity of capital accumulation, as housing has both a use and an exchange value (Duncan 1981: 105).

Until the early twentieth century the overwhelming majority of the population, including the middle classes, in all three countries rented privately. In the UK home ownership became a symbol of social status, and of upward social mobility for the middle classes during the Industrial Revolution (Cannadine 1998; Fishman 1987). In countries like Canada and the USA the desire to own property has been especially strong among first generation immigrants eager to put down roots in the New World (Harris and Pratt 1993: 286). But the desire to own property among those born in Canada and the USA is almost as strong as among newcomers. In one sense this represents a search for social prestige (ibid., Fishman 1987).

Middle-class residences in the urban cores of Anglo–American cities historically were also spaces of enterprise, with some female involvement in family businesses (Fishman 1987). Davidoff and Hall (1987) have shown how gender roles and locational preferences changed over time with industrialisation and suburbanisation. The twin separations – of gender roles and of homes from workplaces were an important feature of industrialisation in all three countries (Fishman 1987; Harris and Pratt 1993). The 'streetcar' suburbs presented a vision of a community based on the primacy of private property and the individual nuclear family in which a middle-class woman's main role as wife and mother was to create a haven from the bustle of public life (Fishman 1987).

It was not until the inter-war years that home ownership became the dominant tenure of the middle classes in Canada, the USA and the UK (Fishman 1987; Germain and Rose 2000; Glass 1964). The tenure shift to ownership is linked to the expansion of the number of households embracing 'middle-class' lifestyles, whose male partners were working in the expanding service sector (Dupuis and Thorns 1998: 24). This was when the semi-detached home (the 'semi') came to dominate the suburban landscape in the UK (Hamnett 1999: 56). The lives of these *'nouveaux riches'* new middle classes were the object of ridicule in inter-war Britain, as in the following extracts from Betjeman's poem *Slough*.

Come, friendly bombs, and fall on Slough
It isn't fit for humans now,
There isn't grass to graze a cow
Swarm over, Death!

....

Mess up the mess they call a town –
A house for ninety-seven down
And once a week for half-a-crown
For twenty years.

....

And talk of sports and makes of cars
In various bogus Tudor bars
And daren't look up and see the stars
But belch instead.

John Betjeman (1937)

In the USA and Canada detached properties in a variety of styles (from small cottages to more elaborate houses) are the dominant suburban housing type (Holdsworth and Simon 1993: 188).

For much of the twentieth century, rental tenure, rather than home ownership, remained the norm for middle-class Montrealers, especially among Francophones (Germain and Rose 2000: 173; Steele 1993: 54), so much so that researchers (Choko and Harris 1990 cited in Germain and Rose 2000: 173) conceptualised this phenomenon in terms of a 'local culture of property'. A gradual shift toward a more typical Anglo–American pattern of lifestyle oriented around domesticity and consumer durable goods, which tends to go hand-in-hand with home ownership, only began to take hold in

Quebec in the 1950s as federal government policies led to the diversification of sources of mortgage credit (ibid.). It was consolidated with the Quiet Revolution[1] of the early 1960s, which brought about a general secularisation of society (ibid.). Despite these changes Montreal still has an unusually high level of rented properties (Germain and Rose 2000).

Housing as an investment

Housing has always been expensive in relation to income and house buyers have always had to struggle to buy for the first time. Most households buy their homes with a mortgage loan, using the property as security. Unlike the situation in the USA and the UK,[2] mortgage interest is not tax deductible in Canada, nor are property taxes, as in the USA (Steele 1993: 42). Yet in all three countries renting is viewed as 'throwing money down the drain' whereas home ownership investment, a way of building up capital, but also on death, children or other family members benefit from the investment (Dupuis and Thorns 1998: 32).

The importance of housing inheritance as a source of wealth is clearly shown in such novels as Galsworthy's *Forsyte Saga* or Trollope's Barsetshire series and George Bernard Shaw's first play, *The Widower's House*. Home ownership has thus functioned as a store and source of wealth: the middle classes have 'moved savings under their own roof'. Housing now accounts for 40 per cent of net personal wealth in the UK, for example (Hamnett 1999). Indeed, the US is said to be 'a post-shelter society' because the investment role of housing is perhaps more important than its traditional role of providing a roof over people's heads (Hamnett *et al.* 1991).

In the buoyant housing market of the 1980s in the UK, people were encouraged to think of housing in terms of a 'housing ladder' or a 'property ladder'. This was a device used by building societies and estate agents in particular to emphasise the investment potential of house purchase (Munro and Madigan 1998). The ladder began with a 'starter home'[3] for the first-time buyer, with the implicit expectation of subsequent advancement up successive rungs of the hierarchical property market. As such the image was seen to provide a 'life plan' which appealed to notions of self improvement and future security and which commanded widespread popular support (Saunders 1989). The idea of a housing ladder embraces financial accumulation: each successive step is buoyed by accumulated equity and capital gain acquired in the previous property.

In the British context (as well as in other locations, such as California) the implied hierarchy appears to carry more than a hint of anti-urbanism in its assumption that a suburban/small town, or better still, rural house and garden is always more desirable than inner-city housing. The property hierarchy is closely associated with an archetypal family lifecycle, in which young adults (whether single or as couples) are expected to make their first house purchase in the city close to work, where flats and small houses, starter homes are available. As the household expands, its income rises and/or the value of its property increases, there may be a move to a more definitely suburban, out of town or a small town location. It is at this stage that children's schooling play a key role in choosing a location. Finally, once the family grows up, there may well be a move to a smaller house or a bungalow near the sea, or a flat near to adult children.

This summary is a gross simplification, yet the imagery is powerful because it combines a consensus of tenure and housing choices with a socially sanctioned lifecycle: single person becomes childless couple, young family, adult household, older couples, and finally a single person. Progress along this trajectory may be cited as an indicator of success and a record of self improvement which meshes easily with many of the perceived advantages of home ownership, including:

- the ability to reduce housing costs in old age;
- a form of savings which is seen as credit worthy; and
- the accumulation of an asset which will ultimately be 'something to leave to the children'.

This idea of progression within the owner-occupied sector has been deeply embedded in the British middle-class psyche for several generations.

Condominiums in North America represent another route into property ownership. Condominiums have a longer history in the USA than elsewhere, but were introduced by provincial legislation in Canada in the 1960s and 1970s. They have broadened the range of home ownership options. Condominiums provide a package of property rights through a legal arrangement that makes it possible for an individual to own a dwelling without exclusive ownership of the land on which the structure is built. Together with the dwelling, each resident of the condominium jointly owns a proportionate share of the common elements, such as driveways, recreational facilities, parking and storage, etc. (Hulchanski 1993: 65). Condominiums are an ownership form not a housing form and can be detached properties, apartment blocks or 'plexes (see p.61), and are especially important in Toronto, Vancouver and Montreal, including in downtown locations particularly favoured by childless professional solos and couples (Germain and Rose 2000). Condominiums have played an important part in the revival of Canadian inner-city areas (see p.61).

The process of residential decision-making

The remaining part of this chapter highlights residential decision-making for dual career households, first by examining the decision-making process, and subsequently the outcome, the preferred residential locations. Both residential decisions and housing search are intrinsically tied to the household, how it functions as a decision-making unit, and the weighting of power relations with it (see Chapter 2). The choice of place of residence profoundly affects the careers of both partners. Residential migration is a highly disruptive process for all those involved, particularly in moves involving considerable distances – it disrupts and fragments a household's social space (Seavers 1999: 151).

Moving home for homeowners is an immensely important operation involving serious amounts of debate and discussion for most couples before they eventually reach a decision to move and buy a given property. Moving home involves a series of decisions:

- the decision to search for a new house;
- the area searched;
- the type of settlement;
- type of property considered;
- size of property;
- internal layout of the property; and
- external general appearance/style of property (ibid.).

The dominance and relative influence of the partners can change at different stages in the migration process, and it is necessary to distinguish between:

- the impetus to move and the actual decision to move; and
- the general destination area and the specific housing search area (ibid.).

Each partner enters the decision process with an image of the desired outcome of the decision, as they perceive it. The areas of disagreement will require the couple to discuss, negotiate, and undertake a series of trade-offs to develop a relative preference or concede their preferences. The impetus to move is often linked to the job market and the pursuit of a career, with residential location strongly related to the male partner's pattern of commuting, while the job search of the female partner is constrained spatially by the residence as a point of origin (Hanson and Pratt 1995; see Chapter 2).

When asked to describe the decision-making process relating to major decisions, especially infrequent lifestyle decisions like house purchase, one man commented, 'I don't actually feel conscious that there is a sort of process by which we sit down and negotiate.' While one woman said, 'I'm sure in many relationships there is always a little bit more of one person in any decision than the other'.

While the choice of general house location tended to be determined by the 'lead' career (see Chapter 2), when it came to the specific house, this tended to be a joint decision. For example, one man said, 'house buying is a bit different because you've both got to see it'. Another man said, 'we were both looking and then tried to make a joint decision on what we wanted.'

Hence, the process of residential decision-making for dual career households involves negotiation and compromise.

Owning a home and having a career with a degree of spatial mobility can be problematic, especially when an inter-regional or international move occurs. A number of couples talked about the problems of having properties to sell some distance away from their new base. Moreover being offered a job where properties are expensive can result in jobs being turned down (as is the case with London, where the highest property prices in the UK are found or for Vancouver in Canada) especially for public sector professionals with lower pay scales than those in the private sector.

Housing as an expression of cultural values and identity

While the actual decision-making process relating to the choice of property tends to be a joint decision, the acquisition of the property is the result of the combined incomes of the two partners and can be used as an expression of the household's cultural capital (Bourdieu 1986; Schoenberger 1997). The preferred living spaces for these households can be examined at the level of the locale. The presence/absence of children within the household and the age of the children also have important implications for residential preferences. The neighbourhood/locality chosen – whether for rented or bought properties – often reflects the desire for a particular kind of living space (Gregson and Lowe 1994a: 226), a space which extends beyond the house itself out into the surrounding environment, 'a civilised retreat' and a site for anchoring 'middle-class' identities (Cloke and Thrift 1990).

As was noted earlier, for the middle classes the home is also an important element of social display and a focus for the expression of cultural values and identity. This role is particularly important for the middle classes, not least because of its function as an indicator of social status and distinction (Bourdieu 1986; Hamnett 1999: 61). Rossi (1980) has argued that there are five ways in which housing can be viewed:

- housing as interior living space: floor space, number of rooms/bedrooms;
- housing as amenities: in that it provides both shelter and amenities;
- housing as location: the position of the housing space;
- housing and associated public services: some services are delivered on an area basis, such as state-funded education, so schools with good examination results can boost an area's attractiveness to couples with school-age children;
- housing context externalities: social and physical contexts in which the housing unit.

In Canada, the UK and USA it is the suburban house that has become an expression of a cultural characteristic, of individualism – a tangible marker of personal achievement, of upward social mobility for the expanded middle classes in the twentieth century (Pratt 1981). The suburb is a thing apart, a human habitat wholly dependent on the city's prior and adjacent existence (Evenden and Walker 1993: 234). It provides the dominant setting for the rearing of families, the learning of property and political relations. It has also been described as a phase in the circulation of capital (ibid.). Suburbia – the unspoiled synthesis of city and countryside – is more than a collection of residential buildings, it is viewed as a partial paradise, a 'refuge' from threatening elements in the city. Suburbia is defined by what it includes, that is middle-class residences and residential space, but also by what it excludes such as industry, most commerce, and lower-class residents (except for servants). The social and economic characteristics are all expressed in design through a suburban tradition of both residential and landscape architecture which are derived from the English concept of the 'picturesque' – the tree-lined avenues of owner-occupied detached or semi-detached villas with gardens.

Feminists believe that housing reflects, and helps to shape, gender relations in society as a whole. Although these relations have been changing in recent years, they are still marked by strong elements of patriarchy. From this point of view, the suburban home, often described as a 'dream home' in all three countries, is viewed with some skepticism. Although such a place may be comfortable to return to after a day in an office or factory, it is a boring, isolated work environment for the (typically female) homemaker (Dyck 1990; England 1996).

Middle-class élites in both Canada and the USA have traditionally shown deference to European styles and precedents, as is the case with Beacon Hill, Boston (Fishman 1987). In Canada, for example this has included an eclecticism of classical styles related primarily to Britain, especially Gothic and Italian designs in the late nineteenth century. While in the first half of the twentieth century the Tudor style mansion occupied a privileged position (Plate 4.1). One of the primary reasons for the influence of British precedents in Canada was the strong personal sentiment of élite members for their native or ancestral lands (Ley 1993a: 216). These housing styles are found in Westmount, Montreal, Forest Hill, Toronto and Shaughnessy, Vancouver, but these styles now form residuals as the areas have been reshaped. In west Vancouver newer élites have made their mark on the exclusive suburb in the post-war era with more modern designs, including ranch-style homes and post and beam structures in the International Style (Ley 1993a: 220; Plate 4.2). A large number of west Vancouver households are immigrants, almost all have moved from elsewhere, and upward social mobility is a common trait (ibid.).

In the USA, the UK and Anglo-Canada middle-class preferences were for individual or semi-detached properties, but in the streetcar suburbs of

Plate 4.1 Tudor-style mansion, Vancouver. Photograph: I. Hardill

Plate 4.2 International-style bungalow, Vancouver. Photograph: I. Hardill

Plate 4.3 'plex apartment housing, Outremont, Montreal. Photograph: I. Hardill

Montreal, Quebec such as in Outremont and Mile End, middle-class properties were largely apartments in the stacked 'plex format (Germain and Rose 2000; see Plate 4.3). The 'plex consisted of a duplex with apartments on two floors, a triplex with apartments on three floors, etc. (ibid. 173). Some of the upper-middle-class apartments were extremely large, up to ten rooms, of a similar size to a substantial villa. The typical pattern in the Montreal case was that the owner-landlord lived at street level and rented the other units (one or two on floors above). While apartment living in 'plexes and purpose built, for rent, apartment buildings of three to five storeys, dominate bourgeois Outremont, there are also substantial villas too. In the Anglo Montreal suburb of Westmount the dominant property type was substantial villas.

The geographical literature on the residential location strategies of employed couples has suggested that within the broad category of the 'dual career' couple there exists diversity of gender practices and structures of consumption and in the choice of living spaces (Germain and Rose 2000: 204; Hardill *et al.*, 1997a). Using Canadian data for 1990, Germain and Rose (2000: 204) have shown that in Montreal it is younger, often childless, dual career households who tend to live in the inner city. They work in the parapublic, cultural, artistic and communications sectors, some being employed on fixed term contracts, self employed or employed part-time (ibid.). Also within the inner city are more affluent childless dual career households who live in the traditional élite areas or in redeveloped infill sectors (ibid.). Those dual career households with children tend to gravitate to the suburbs given the high cost of single family housing in the inner city, and the scarcity of neighbourhood green spaces.

While there is diversity in choice of living spaces for dual career households, suburban development and life have emerged as dominant experiences in Canada, the USA and the UK. Although today the suburbs display a wide spectrum of social conditions and are not associated exclusively with an emergent middle class, they remain the favoured residential location of dual career households with dependent children. The four bedroom detached 'executive' house has now replaced the inter-war suburban 'semi' as the typical home for the enlarged middle classes in the UK. Housing styles in North America and the UK have thus merged, but properties in North America tend to be larger. Most suburban dual career households do not live in élite suburbs however, but in private housing estates in peripheral locations (see below; McDowell 1997). They live there for only a few years before relocating for the next job. They inhabit a different type of suburb, 'more economically independent from the urban core, but not really a town or village either; a place that springs into life with a developer's wand, flourishes ... no one in them becomes a long-term witness to another person's life' (Sennett 1998: 20).

A strong preference for properties in accessible semi-rural locations in central England has emerged (such as near to a motorway hub, or to an airport) thereby providing access to several labour markets, for both childless households as well as for those with dependent children, with both partners having access to a car (Green 1997; Hardill *et al.* 1997). The choice of such locations represents a trade-off between a longer commute and migration, recognising that migration can adversely affect the career of the trailing spouse partner. Such accessible locations are not a viable option in countries of continental proportions, such as the USA and Canada, where few such nodes with access to several labour markets exist. In North America for most careerists daily commuting is not a feasible trade-off with migration. Instead there appears to be a greater acceptance of long-term 'commuter marriage' in order to prioritise two careers (see Chapter 2). Daily commuting from cheaper housing areas in the Central Valley of California to Silicon Valley or San Francisco is now common. Some even commute weekly by air from Nevada and Oregon because of lifestyle preferences for the household and lower housing costs. They may be office-based four days a week. Living near to an airport is also important for sales people who do not have a fixed place of work.

In North America during the latter part of the twentieth century, the outer edges of the major cities experienced astonishing rates of subdivision and development, much of it taking the form of 'sprawl' (Evenden and Walker 1993: 235). Urban sprawl refers to spatially unco-ordinated processes of development; an anarchic landscape made possible by increasingly widespread ownership and use of motor vehicles (McDonald and McMillen 2000: 135). Writing about the USA, Joel Garreau (1991) has argued that 'edge cities' represent the crucible of America's urban future. The classic location for contemporary edge cities is at the intersection of an urban 'beltway' and a hub and spoke lateral road, where 'community' is scarce. In the USA the 'technoburb' peripheral zone has emerged as a viable socio-economic unit, as along the highway growth corridors are shopping malls, industrial parks, and campus-like office complexes, hospitals, schools, and a full range of housing types. The archetypal technoburbs are Silicon Valley or Route 128 in Massachusetts. With the technoburbs we can see the re-establishment of linkages between work and home, as they contain both home and work within a single decentralised environment (Fishman 1987). It has been argued that the technoburb is now the true centre of American society (ibid., 198); a standardised and simplified sprawl, consuming time and space, destroying the natural landscape.

The process of resettlement of the countryside, or more specifically 'the city's countryside' is occurring in all the three countries (Bryant and Coppock 1991: 209). Some 50 per cent of British people would like to live in the countryside (Countryside Agency 2000). The perceived quality of the rural environment is one of the most important factors in the appeal of rural areas as places to live and surveys reveal that rural dwellers are more content than urban dwellers. In-migration to rural areas may be conceptualised as part of a wider 'counterurbanisation cascade' or the suburbanisation of the countryside, and reflects the attractiveness of rural environments (Champion *et al.* 1998). The reasons for migrating to rural areas have been linked to middle-class migration and rural gentrification, but the service class connection has also been critiqued (Phillips 1998). Rural inmigrants in all three countries include households with dependent children seeking what they perceive to be a better environment (lower crime levels and less pollution) an 'idyllic' alternative to the perceived problems of urban areas coupled with the availability of better schools (Phillips 1998). They favour accessible rural and semi-rural environments within the 'urban field' (Dahms 1998), with residents undertaking long-distance daily or weekly commutes (Green *et al.* 1999; Hardill 1998; Momsen 1984). Their lifestyle demands access to a good transport infrastructure, in particular the motorway network or proximity to an airport.

Janet Henshall Momsen's (1984) study of in-migration to the Calgary-Edmonton corridor in Canada illustrates this trend well. The surveyed migrants tended to be two wage households, with dependent children, who moved because of a somewhat 'utopian' view of the countryside and country life (ibid. 170). Selection of a rural area for in-migration is usually the result of a complex interplay of factors, but the search for the 'rural idyll' (Momsen 1984: 166) is a key unifying factor. This is well exemplified by the comments of a young rural in-migrant family to the East Midlands, UK,

> 'we have always liked country life and because we've got children that was another reason for us to move to the countryside'.

In a study of migration to rural areas 50km south of Ottawa (Skaburskis and Fullerton 1998) similar findings emerged. Most (80 per cent of those surveyed) indicated that the quality of the environment, the rural setting, the size of property and the price were the key factors in their decision to move to the countryside. While their average commute time had almost doubled, it stayed within the one hour norm. While the 'peace and quietness' was valued, the commute time and commute costs were sources of dissatisfaction (ibid. 253–4).

The move to a rural area in all three countries resulted in intensive car use (Momsen 1984), as one British rural in-migrant commented:

> 'we have to run two cars. We have no option on that. We actually live out in the sticks. Other than that nothing really bugs me'.

And one British male college lecturer, who lives in a small village said:

> 'the big thing about living in the countryside is you spend your whole life in the car. You have the idea of this healthy open air type of life and it isn't like that. I've never driven so much in my life'.

Thrift and Leyshon (1991 cited in Hamnett 1999) have argued that the boom in the English country house market in the mid to late 1980s stemmed from the rapid rise in income and wealth, particularly by those working in the City (of London) and financial services. Some who had made money in the City or via Thatcherite entrepreneurialism also had a desire to buy into the traditional 'country gentry' lifestyle. Outside the country house market, people who are successful in business are likely to buy a house to symbolise their success in its size and distinctive attributes (Hamnett 1999).

In addition to in-migration second and holiday home purchases are serving to fuel the private housing market in the UK. Fifty-eight per cent of England's second homes are in rural districts; even though the percentage of homes which are second homes remains low (0.9 per cent in rural areas compared to 0.4 per cent nationally). In some areas, however, the proportion is significant, such as 20 per cent in the Isles of Scilly and 16 per cent in

Windermere (Cabinet Office 1999). Holiday homes, rented to others, and occasionally used by the owners, account for a further 300,000 properties (Hetherington 1999). In some localities, especially in north Wales and Scotland, second homes bought by English people have caused much resentment and even violence.

A minority of dual career households make the decision to opt out of the 'rat race'. This represents an attempt to 'take control' of their careers and therefore their lifestyles, often through moving from employee status to an entrepreneurial strategy of self-employment (Sennett 1998) or by accepting job contracts that create space for a chosen lifestyle. For some this is achieved by using their skills and expertise and social contact network from their previous job, as Bill and Betty[4] did. The 'opting out' may be accompanied by choosing a residence in a rural area, sometimes a rather remote rural area but with good virtual links through the use of information and communication technologies (ICTs) (see Chapter 5; Brooks 2000; Norton 1999; Wilkinson 1999).

For others, high-pressure careers and large incomes are traded in for a less frantic and more creative life by becoming self-employed (Knowsley 1999). Graduates from a one year course run by the English Gardening School in London, for example, used to be people in their fifties and sixties, today they are people in their thirties (ibid.). Recent graduates include former City (of London) high-fliers, who gave up six-figure salaries and high stress levels for a more healthy work environment and a job which offers them a chance to be more creative (ibid.). As one 37-year-old graduate from the English Gardening School said, 'towards the end of my time in the City, all I did in the few hours that I saw [my family] was moan. I was very unhappy. I would leave home by 6am and often not return until 2am the following morning. I did make a lot of money. [Now] I have a better life, [even though] I will be working for around £2 to £3 an hour – at least initially' (ibid. 19).

Recent work by David Ley (1993; 1993a) has highlighted the fact that some of the old élite areas in the inner city and in the inner suburbs have maintained their high rank throughout the twentieth century. While wealth has tended to flee the central city in the USA, this is not the case in Canadian and some British cities. Some of the desirable élite middle-class residential areas, such as Hampstead, London; the Park, Nottingham; Rosedale, Toronto; Westmount, Outremont and Mount Royal, Montreal; and Shaughnessy in west Vancouver remain desirable, exclusive and expensive (Ley 1993a: 215). Not only have they survived, and sometimes been expanded by gentrification, they are beyond the means of most members of the expanded middle class. The Park, Nottingham, which is a private estate that was originally part of the demesne of the famous Nottingham Castle provides a good example. Its construction began in the early nineteenth century, the layout being carefully planned (Hardill *et al.* 2001). It was a prestigious area with large houses and mansions in small but secluded grounds. In its heyday the Park housed around 4,000 people in some 650 houses. These were people of stature – lace barons, professional people, manufacturers, merchants and industrialists. Among the better known of the Park's residents were Jesse Boot (chemist/retailer), A. J. Mundella (politician/reformer) and John Player (tobacco baron). There has been much subsequent in-filling of grounds and more recently subdivision of the larger and smaller houses. A number, especially those built in the 1960s, are badly designed and uncharacteristic buildings. The Park still attempts to retain its exclusivity by closing the gates once a year. Access to the Estate from Castle Boulevard can only be gained by means of a residents' card. It has its own street cleaners and security personnel paid for by the residents through a special levy. New residents must sign and adhere to special regulations. There is a management committee, estate manager and office. Though some exclusive sporting facilities exist shops, pubs, businesses and bus services are lacking.

Housing styles in élite suburbs in Canadian cities, especially Toronto and Vancouver, have been transformed in the last decade or so as skilled migrants from Hong Kong, with high profits from real estate have acquired properties in these suburbs, often buying them outright (Smart and Smart 1993). The negotiated handover of Hong Kong to China in 1997 raised concern about whether the economic miracle could be maintained. Those who could leave, namely the affluent, skilled and educated have seen Canada as a 'safe haven' or 'political insurance' (ibid., 37). They have been able to select élite locations in the most expensive parts of Canadian cities, often areas which previously were neighbourhoods dominated by Anglo upper-middle-class or upper-class populations; they have purchased properties, often outright and redeveloped them. The original property and large garden have been replaced with what are often referred to as 'monster homes' (see Plate 4.4, p.60). Areas like Kerrisdale (former Anglo middle- class neighbourhood in Vancouver) or Willowdale (a Toronto upper-middle-class Anglo neighbourhood) have been changed as huge houses filling the lot have been constructed, covering the former manicured gardens, and the Hong Kong migrants have 'paved paradise' (ibid. 36). Some of these households have a dual-household 'astronaut' arrangement, in which one of the spouses returns to work and lives in Hong Kong and the other spouse and children remain in Canada (for a fuller discussion see Chapter 6).

Since the 1970s some British and North American cities have re-emerged as cultural centres, with some neighbourhoods undergoing transformation through gentrification, that is, the movement of wealthier households into districts formerly occupied by less affluent residents (Ley 1993a: 228). Cultural and artistic facilities and activities are also deeply imbricated in the dynamics of gentrification (Germain and Rose 2000; Ley 1996; Smith 1996). The extensive literature on gentrification has shown that a downtown concentration of advanced tertiary jobs is a necessary but not a sufficient condition for new middle-class settlement in inner-city neighbourhoods (ibid.).

Plate 4.4 'Monster homes', Vancouver. Photograph: I. Hardill

The origins of gentrification around 1970 coincided with the demographic bulge of the baby boom entering the housing market, and there is evidence that an inner-city address represented a counter-cultural act, or at least an expression of an alternative cultural politics, for a fragment of the new class. This fragment was often well educated and rich in cultural capital, pre-professionals or people at an early stage in their career as a social or cultural professional, and may not have been well endowed with economic capital (Ley 1996). The perception of this cultural new class counterpoised the world of the conformist suburbs unfavourably against the more diverse opportunities of the inner city (Ley 1996a: 25).

Gentrification is a complex phenomenon, and it can involve either renovation or redevelopment of housing stock (Ley 1996). This 'new urbanism' in North America (in Canadian and some US cities) has been inspired by the living example of the old city neighbourhoods of Montreal with their 'plex housing, narrow streets and street corner businesses (Germain and Rose 2000: 167). The gentrification submarket has also diffused beyond the young urban professional to include a growing number of families with children as well as a significant share of 'empty nest' households who particularly favour new city centre condominium structures (Ley 1996a: 24; Plate 4.5). Gentrification is creating new landscapes of privilege in the inner city, largely the result of the growth in professional, managerial, administrative and technical occupations, who have opted to live downtown (Ley 1996a: 18; Plate 4.5).

Plate 4.5 New élites: False Creek, Vancouver. Photograph: I. Hardill

Inner-city living is seen to offer an alternative lifestyle as expressed through renovated housing, the conversion of existing buildings or the construction of condominiums, with careerists living in penthouses, New York-style lofts, town houses, or waterside apartments (see Plate 4.6, p.62). These properties represent a self-consciously urban architecture and lifestyle in many ways the antithesis of the suburban ideal, and are associated with the rise of the two career household, especially young childless couples but also solo professionals (Ley 1996; Smith 1996). These household members tend to invest heavily in their careers, with activities such as work/leisure/home becoming 'blurred'. Leisure thus becomes an activity where one networks, whether at the gym or at the restaurant (McDowell 1997). The type of accommodation chosen is also a lifestyle statement and expression of status and 'belonging'.

In Toronto and Montreal, the process began in downtown areas adjacent to élite neighbourhoods (Ley 1993: 229–30). In Montreal the earliest gentrification in the city involved large Victorian townhouses on the downtown fringes with exceptional architecture which had deteriorated. These were 'reclaimed' by professionals in the 1970s (Germain and Rose 2000: 199). The period 1981–6 when the City's renovation-stimulation programme was in full swing saw transformation, especially of the Plateau Mont-Royal area, by highly educated but not necessarily wealthy professionals, a number of whom are self-employed and work out of their own homes (ibid. 188). Wealthy dual career couples also live in new condominiums on Île des Soeurs, an island in the St Lawrence (ibid.). In Vancouver more generally,

Plate 4.6 New élites: Vieille Ville, Montreal. Photograph: I. Hardill

environmental amenity has played a significant role in directing the course of gentrification (Ley 1996a: 23). Middle-class investment in the city has tended to take the form of condominium redevelopment rather than the renovation of heritage properties. The transformation of Vancouver's Expo lands into a highly desirable downtown living space was facilitated by Hong Kong capital in the late 1980s (Mitchell 1995). In Kitsalano, Vancouver for example the construction of new condominiums has helped transform Fourth Avenue into an area with trendy shops, restaurants, and speciality shops (Ley 1993: 231).

Tim Butler (2001) has shown that gentrification is a diverse process in the UK too. He has looked at how a number of areas of London are being transformed by middle-class gentrification. Each of these areas has a different 'style' which he argues represents differences within the middle class which become embodied in the different characteristics of these areas and the way in which they socialise incoming members of the middle class. He suggests that the ways in which the middle classes are creating neighbourhoods can be distinguished by the variations in their deployment of their stocks of cultural, economic and social capital (Bourdieu 1986). The outcome is strongly spatial, with strong 'local cultures' emerging. Middle-class people in London, he argues, are exercising choice about where to live, partly determined by economic factors such as housing market constraints but also by identification with neighbourhoods where they perceive 'people like us' to live. In the case of Barnsbury (where Prime Minister Tony Blair used to live) and Battersea, for example, he argues, there has been a wholesale accommodation to global culture largely by its London and City-based functionaries, resulting in these places also developing their own distinct form of metropolitan cultural/consumption infrastructure. The gentrification of London received publicity through the film *Notting Hill*, starring Hugh Grant and Julia Roberts. The film certainly had a positive effect on property prices!

Conclusion

This chapter has explored what it is to be a manager and a professional in terms of status and success, with particular reference to the purchase of residential property. Since the Industrial Revolution property has taken on added significance for the middle classes. In today's post-shelter society however, residential property has multiple meanings for the middle classes in general and dual career households in particular. While residential property is viewed as an investment, as an asset, it also has a cultural significance, being a form of expression of status and success. Housing thus has a great symbolic meaning to many dual career households. Personal and household mobility is shaped to some extent by the property market.

5 The blurring of boundaries between home and work

Introduction

As was noted in Chapter 4, since the Industrial Revolution for the middle classes 'work' and 'home life'[1] has taken place in different domains in advanced capitalist economies such as Canada, the USA and the UK. The home is both a physical location – as a bounded and clearly demarcated space – and a psychological concept. Often presented in positive terms of warmth, as security, as a haven from the pressures of work and public life (Bowlby et al. 1997: 343). Houses are assumed to become homes because they provide and become the environment within which family relationships – close, private and intimate – are located. While it is true that non-family households also have homes, a crucial understanding of home is a notion of a place of origin, a place of belonging, a place to which to return, a place where children are/will/may be reared. Thus the white, Western ideology is of a home as a physical entity, and a set of social, economic and sexual relations (Peck 1996: 37–40).

The 'traditional' view of the home is of 'family relationships': made up of a heterosexual couple, married or co-habiting, with children or other relatives. But the home can also be regarded as a site of oppression because of patriarchal practices, such as when women's labour is expropriated by a male partner within the household; by male violence and patriarchal relations in sexuality as well as patriarchy in culture (Walby 1989: 214).

A second strand of research being undertaken by feminists is looking at 'when' and 'where' work (for economic relations) is undertaken, and its impact on the 'home' (Chapters 1 and 2; Hanson and Pratt 1995; Jarvis et al. 2001). While women are more 'visible' in work for economic relations, they participate in the labour market on a very different basis to men. As was noted in Chapter 2, the number of hours devoted to paid work are increasing, and dual career households are 'time poor', but cash rich (when compared to other household types) (Doyle 2000; Hills 1995). Though some dual career households have reduced the hours they devote to unpaid work by commodification of domestic tasks, many female partners still make a heavy temporal commitment to 'home' in addition to their paid job

(Chapter 2). In the remaining part of this chapter we explore the ways in which paid work patterns are changing through the rise in the number of hours devoted to paid work, the use of Information and Communication Technologies (ICTs), changing commuting patterns, the diverse spaces of paid work, and the health-related impacts of working all hours.

Working all hours: the lived reality

'Working 9–5' is not the way most people make a living today. We now work in a flexi-time, flexi-place world. With increased dependence on instant electronic communication the timing of the working day may be shaped by working times in another time zone: for example, the stock market in Vancouver follows New York trading times. The economic power of the hour is diminishing with the evolution of a post-industrial economy because 'knowledge work' does not easily fit into the '9–5' structure; '24–7', global economies require new kinds of working. Moreover, knowledge work is not as linear as manual; it is erratic, inconsistent and highly personal. Individual workers perform better at different times of the day and under different circumstances (Doyle 2000).

Since the advent of the Industrial Revolution the story of working time has been one of a steady decline in working hours. In most European Union countries this downward trend continues to this day, but in the UK, Canada and the USA this is not the case. The length of the standard working week is not increasing; rather the increase in total hours devoted to paid work is the result of overtime, especially unpaid overtime. It has recently been argued that payroll taxes and fixed benefit costs in Canada work in such a way that employers whose employees work extra hours are rewarded with lower payroll taxes and reduced benefit costs (O'Hara 2001: A15). An analysis of working hours shows that professional women in the UK, for example, put in an average ten hours unpaid overtime a week, up from four hours a decade ago; for men the rise has been from five to seven (Harkness 1999). An analysis of the British Labour Force Survey (winter 2000–01) reveals that 38 per cent of male managers worked over 48 hours per week, while 32 per cent of women professionals worked over 48 hours (cited in Doyle and Reeves 2001: 13).

This lengthening of the working week is in contrast to trends in France, where the European Union Working Time Directive (for a 35-hour working week; see Figure 5.1, p.66) has been implemented under the Aubry law. The Aubry law set working hours to an annual ceiling of 1,600 hours, and it is broadly left up to individuals and firms how to distribute those hours, but for most, the annual limit translates to a 35-hour week. The law came into force in January 2000, and to date (June 2001) covers 47 per cent of full-time employees (Futures 2001). The introduction of the Aubry law appears to have actually increased labour market flexibility in France, but restructuring public

- A limit of an average of 48 hours a week which a worker can be required to do. Average worked over a 17 week period.
- A limit of an average of 8 hours work in 24 hours which nightworkers can be required to work.
- A right for nightworkers to receive free health assessments.
- A right to 11 hours rest a day.
- A right to a day off each week or 2 days in a 14 day period or a 48 hour period in a 14 day period.
- A right to an in-work rest break if the working day is longer than 6 hours.
- A right to four weeks paid leave per year.

Figure 5.1 The Working Time Regulations 1998
Source: Department of Trade and Industry

sector working patterns is certainly not proceeding smoothly (Le Snesup 2001). Each company signing up to the 35-hour week has had to negotiate an agreement with its employees about the implementation of the accord. Since the introduction of the accord could not be accompanied by a wage cut, companies have used this to leverage from their employees greater flexibility in terms of shift working, weekend working, holidays, etc.

It is not just full-time workers who work more than their contracted hours, those on part-time contracts also do. A report by the Industrial Society (Doyle and Reeves 2001) points out that people in the UK (full-time and part-time) are working above contracted hours because it is the only way they can get things done, but also that for 47 per cent of those surveyed it was their 'choice'. They go on to argue that what is critical is the ability to control where and when paid work is undertaken, such as opting to travel to work outside congested rush hours, as well as the quality of work. Workers who have autonomy over their diaries are termed 'time sovereign' (Doyle and Reeves 2001). Feelings of being overworked tend to be associated with those workers who feel they do not have the flexibility to control their work time and manage personal and family responsibilities (Doyle and Reeves 2001: 24; Reeves 2001). Others have written about the practice of 'presentism', the need to be visible in the workplace, and the pressure of not being the last employee to arrive on a morning, or be the first to leave at the end of the working day (Doyle and Reeves 2001). Fear of the future, such as not being on the next list for redundancy, or the fact that 'commitment' is a pre-requisite for promotion, results in workers giving more and more time to work (Dean 1998). Some of these negative feelings towards work can be eradicated by reverse telecommuting, which refers to the process of bringing personal work into the office (making photocopies, making say a dental appointment, booking a holiday) (Reeves 2001). Much more is written about the other side of the coin, of flexibility being a one-way street, with the employee always giving, such as completing a report at home.

Some recent research with British women has focused on how they balance 'work' and 'home' with their own leisure needs (Gilroy 1999) and on the blurring of work and leisure (McDowell 1997). Household negotiation and playing out the dynamics of daily life affect one's partners' leisure time. Time is a crucial commodity as far as leisure is concerned, especially for women in dual career households. Being in paid employment in many ways enables leisure, not just in so far so it may increase the amount of money that can be spent on leisure. Todd (cited in Utley 1998) has identified four main forms of leisure: leisure as an extension of work – conferences, drinking and eating with colleagues and networking; partner leisure, such as going out for a meal, or simply synchronising diaries to relax together; family leisure, the most significant category, which consists of being with children – a blurring of leisure and the tasks of social reproduction; personal leisure, a poor fourth, which includes health and fitness, arts events, etc.

The blurring of work and leisure spaces has also been emphasised in a study of a very different professional group – merchant bankers in the City of London (McDowell 1997). Indeed in the TV programme *Ally McBeal*, work spills over into social/leisure time as the workforce relocates from the office to the bar in the same building.

I now draw on interview material to illustrate this discussion. A relatively common complaint was that hours worked were excessive, for example, Simon[2] felt he 'put in more hours than I should ... flexibility means for you to work an extra hour as opposed to the other way ... we have baked beans on Tuesday, Wednesday, Thursday and Friday if we are having a bad week!'

He went on to comment on 'presentism' saying, 'you must be there; it doesn't matter that you are half-asleep. You have to be seen there; a lot of people we know sit in their offices reading their newspapers for an hour between 5 and 6'. As with most others expressing similar sentiments, he felt that 'commitment' (as measured by working long hours) was an essential element in long-term career progression. Hence, although he recognised that long hours were damaging, he felt unable to reduce his hours to remedy the situation; he was caught in pursuing a temporal strategy to achieve an individual work career.

As was discussed in Chapter 2, the partners in dual career households tend to work above contracted hours, but also tend to have some flexibility regarding when work is undertaken. Some (but not all) respondents had some flexibility about where work could be undertaken (see above). As Ken[3] notes, 'How I design my diary is entirely down to me, how I work, how I operate, if I wanted to base myself at home for the whole of next week I could ... if I screwed it up I would be looking for the next job'.

Sarah[4] commented about 'academic flexibility', 'I now do 3.5 days a week [as a university researcher] but flexible [she varies when and where she works], which is the beauty of working for an academic establishment'.

Some individuals could only work at their workplace; those in medical occupations were the most obvious examples amongst the case study households, but personnel staff also tended to be constrained similarly. Jason[5] is also spatially constrained as a scientist, 'I have to go in five days a week because the bacteria I work with needs to be looked at every day just about'. Aside from this, attitudes and practices about where work should be undertaken occupied a wide spectrum. The majority of individuals in the high-level non-manual occupations represented in the dual career households, however, generally accepted that some porosity of the work-home boundary was beneficial for both the employer and employee.

Some individuals and households had made an explicit decision to separate home and work spatially as much as possible (for further details see Green 1997; Hardill *et al.* 1997a). While some used the journey home to 'switch off' from work-related tasks, others (especially rail commuters) used the journey as an 'extension of the office' to undertake tasks which might otherwise have encroached on the home. Samantha[6] said, 'some of the things I could do at home but I wouldn't like to because I don't like anything at home related to work. I like to be able to switch off, so I wouldn't like an office at home'.

Being 'available' for work for some meant that evenings and weekends were no longer sacrosanct. Thus Suzanne[7] said, 'there are things you can't take home. I need to go back and do them; things I don't finish I can go back in in the evening or early morning [to do]'. Employers also tend to assume that it is not a problem for business trips to include a weekend, not to mention short assignments away from home. Thus over a quarter of the surveyed British households had lived as a commuter couple since being together because of a short-term assignment in the UK or abroad[8] (for a further quarter it was the result of co-ordinating relocation because of a new job).

Four of the British interviewees mentioned that they had failed marriages, and two felt that work had been partly to blame. They had prioritised their careers to the detriment of the relationship; they had invested too much time in the career rather than time 'being there' for the relationship. Joanne,[9] for example, talked about her career and a failed marriage. She had been working very long hours in a hospital and then, 'my [first] son died and about 3 months after my [first] husband went off with his secretary, I had to review what I was doing'. She then decided to switch to general practice to try and 'have a life'.

Telework

Increasingly employers are reducing real-estate costs by closing or scaling down costly office facilities, and equipping employees with portable notebook computers and telecommunications equipment to enable them work from home, in cars or in some cases, at the client's premises. Other employers, especially in the USA, are eliminating dedicated office space as an entitlement, moving towards the assignment of office facilities as a resource to be allocated on an as-needed basis through programmes known variously as 'hotelling', 'just-in-time' offices or 'non-territorial' offices. Telecommunications advances have led to a steady increase in the proportion of jobs that can be done outside an office/work setting.

The idea that more widespread home-based teleworking could substitute for travel-to-work and so deliver social advantages including reduced energy consumption, less traffic congestion and fewer polluting emissions from cars has had an enduring attraction since the oil crisis of the early 1970s (Baines 1999). While there is no doubt that information and communication technology is changing the ways that people work, to date there have been no large-scale radical changes on settlement patterns and housing demands (Hardill and Green 2001). Christie and Hepworth (2001) note that:

- the 'information economy' is most fully developed in existing areas of economic development – most notably major cities;
- there is no mass movement towards telecommuting.

Rather than telecommuting on a large scale, the workers of the information economy seem to be living in suburbs and rural areas, and commuting to workplaces in offices in cities (Breheny 1999). The information flows associated with the increased use and penetration of ICTs are bolstering the role of cities as centres of interpretation of information flows (Gillespie *et al.* 2001). Writing about New York, Sennett (1998: 107) notes that while modern communication technologies may well have speeded up the process of collaboration, face-to-face still is the major means of transmission, the 'buzz' at parties, clubs and restaurants is still essential for business.

The dense face-to-face relationships, transactional opportunities and agglomeration economies, and social and cultural resources of cities, remain extremely attractive. Fitzpatrick (1997) has suggested that remote working from self-sufficient farmsteads via the Internet can in no way replace the powerhouses of personal interaction which drive teamwork and creativity. The electronic 'chat room' is an impoverished form of human interaction, and e-mail is an impoverished form of human communication compared with personal relationships.

Even if some commentators believe that the most striking feature of electronic communication is how little geographical diffusion of activities it has produced, it is important not to discount:

- those changes in working arrangements that are taking place; and
- their implications for the individuals concerned, their households and their housing choices.

ICTs in combination with a move to non-territorial offices have revived homeworking, a historic work pattern (see Chapter 4). Moving the workplace into the home carries risks of social isolation and, particularly for female workers, increased work/family stress (Chapter 2). A survey of work-life balance practices in Great Britain conducted in 2000 (Hogarth *et al.* 2000) revealed little evidence of extensive working from home. Of employees in establishments with more than five employees, approximately 80 per cent worked exclusively within the establishment that employed them and 20 per cent worked from home. One third of employees responded that they would like to work from home, at least occasionally. Eighty per cent of employees currently working from home at least occasionally were managers or professionals, and 35 per cent of all managers and professionals currently worked from home to some extent. Hence, the picture emerging is one of many people working from home, or using their home as an office, for at least some of the time, while others work at the workplace in 'normal' working hours, and routinely work at home at other times. Most cases of working from home are on an 'occasional' basis (Hardill and Green 2001).

Counts of homeworkers and teleworkers are fraught with difficulty. It is estimated that about 6.7 per cent of those in employment in Great Britain had 'home-based jobs' in 1991 (Felstead and Jewson 2000: 60–1); in 1996 six per cent of the working population of Canada teleworked (Johnson 1999), while in 1994 an estimated six per cent of the US workforce were telecommuters (ibid.). In a major review of homeworking, Felstead and Jewson (2000) suggested that counts of homeworkers could vary on the basis of:

*   forms of employment that are eligible for inclusion in the count;
*   time thresholds required to trigger inclusion of an individual within a particular count;
*   the location of work – whether a loose interpretation encompassing any paid work connected in some way with the home is included, or a more specific and narrow definition is adopted.

As regards teleworking, some information is available from a major British employer survey of eWorking conducted by IES (2001). eWorking is defined to include any activity that involves the processing of information and its delivery via a telecommunications link that is carried out away from the main premises of an organisation. As outlined in Table 5.1, eWorking may be carried out by an employee of the organisation, or it may involve outsourcing. It may involve people working individually at home, or working together in shared non-domestic premises away from the organisation's main establishment base.

Adopting this broad definition of eWorking, it is estimated that around half of establishments in the European Union are practising some form of eWork. The stereotypical employee teleworker based solely at home is one of the least popular forms of eWork: only 2 per cent of establishments in the European Union employed people to work exclusively from home in this

Table 5.1 Definition and categorisation of eWorking

	In-house	Out-sourced
Individualised	fully home-based working by employees; multi-locational or nomadic working by employees	freelance work
On shared premises (which may or may not be described as a call centre)	remote back offices; work by employees based in telecottages or other non-domestic premises; owned by third parties	business services supplied by independent contractors

Source: IES (2001)

way. Ten per cent of employers used new technologies to support multi-locational working, and 11 per cent used freelancers to deliver work by means of telecommunications. Around seven per cent of employers were found to have a remote back-office outside their own region, and one per cent used telecottages or telecentres as remote bases for employees. Forms of in-house teleworking were found to be outnumbered by eOut-sourcing as a means of carrying out work remotely.

For those involved in working at home, either on a regular or an *ad hoc* basis, the ways in which temporal and spatial boundaries both within the household and around the home separate it from the outside world have important implications at individual and household levels. Felstead and Jewson (2000) characterise different arrangements, and the different degrees of 'blurring' between 'work' and 'home', as follows:

- open – weak external boundaries between the home and the outside world;
- closed – strong external boundaries between the home and the outside world;
- segregation – strong internal boundaries within the home and between the times and places of paid work and domestic life;
- integration – weak internal boundaries within the home and between the times and places of paid work and domestic life.

If an individual using ICT to work exclusively from home is adopted as the 'stereotypical teleworker', teleworking has not 'taken off' in any grand way. The stereotypical teleworker is selfemployed, such as freelance editors, proof readers and translators (Baines 1999). Yet ICT has had profound implications on working arrangements, as the following practices have increased:

- 'partial teleworking' – working at home for part of the time, on a regular, occasional or *ad hoc* basis (sometimes called telecommuters). Such workers tend to be male, with employee status and relatively highly skilled, with high trust relationships and some degree of greater autonomy as to when and where they work. The home is thus the extension of the workplace; diaries are 'monitored' by e-mail.

- hot-desking – when employers provide less office space than required by the full complement of employees, on the basis that not all employees will be in the office every day; (hence employees work at any desk within the office, rather than having their own individual dedicated workspace);
- mobile working – whereby the worker has no fixed base, but rather works out of a car/their home (Graham and Marvin 1996). Increasingly railway carriages and aeroplanes resemble 'offices' with mobile telephones and laptops being used by business travellers and commuters. Amtrak in California has connections for laptops, as do the ferries around Vancouver, British Columbia, to attract and cater for their business travellers. Indeed business travellers are heavily targeted in marketing campaigns, with advertisements stressing that the airline/train operator caters for their 'office' needs (Plate 5.1).

All of these working arrangements involve use of the home for 'work', at least for part of the time. Telework in its broadest sense therefore encompasses many temporal and spatial patterns of work, class positions and forms of employment (Baines 1999; Felstead and Jewson 2000; Green *et al.* 1999; Hardill and Green 2001). Since it recasts the boundaries between

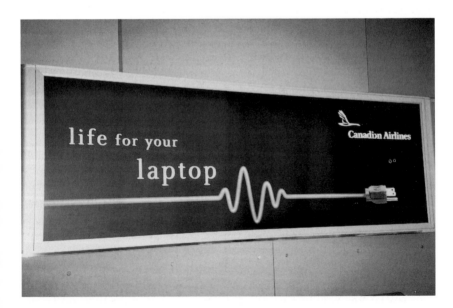

Plate 5.1 Working and travel, Toronto Airport, April 1999

home and work, home-based work has advantages and liabilities for workers trying to balance work and home (especially family responsibilities). By not restricting work to a particular place or time, home-based telework may increase the flexibility of accomplishing work. Research undertaken in Sweden and Canada (Johnson 1999: 122) found that the timing of work tasks by teleworkers was in fact dependent on the time use of others – many interviewees felt that they had to be available for colleagues and clients during work hours, and some also reported over-working in the evenings and on weekends. While telework does free up commuting time (Felstead and Jewson 2000; see p.69) one should not forget that teleworking/telecommuting can also result in childcare issues becoming 'hidden'. Indeed the problem of trying to juggle home and work is often cited as the reason for women managers and professionals becoming self-employed teleworkers (Telecottage Association 1999; BT Working from Home 1999).

The technological linking of home and work was mentioned by a number of employees in the case study households. E-mails have become a means of communication both internally within an organisation and externally. In the UK an average of 22 e-mails a day are received (Walsh 2001). E-mails are replacing hard copy memoranda and correspondence by letter, but e-mails are a source of stress (ibid.). A number of interviewees talked about the growing work load associated with dealing with e-mails. As Hela,[10] a university professor in the USA remarked, ' The way my days [are organised], I stay for the first two hours in a morning at home. I organise my life. For example, my e-mail I don't access e-mail during the day [at work]. E-mail these days carries a lot of work. I get 30 to 40 e-mails a day. The reason I'm entering e-mail in the morning is I'm connecting with Europe because of the six hour difference'. Hela is 'time sovereign', she can control her diary, but also has autonomy to decide where she undertakes some work tasks. For her, the 'organisation of life' means undertaking some work tasks, such as correspondence; checking and replying to e-mails while her correspondent might still be on-line in Europe, a tangible reminder of globalisation. As many of us know, the number of e-mails we receive is increasing; not checking e-mails for a few days leaves one with a long list of unread messages.

Other interviewees talked about telework. Carrie[11] described a former job, 'it was home-based but the office was in London, so occasionally I had to go to London. At home I was based in the upstairs bedroom. It was good to start off with. The actual hours were worse than I had ever had before, because of the regime they had. You worked 8 to 5 without a lunch hour. In the evening you were also expected to work at least 3 to 4 hours'. For Carrie, 'working at home' was qualitatively a very different (and worse) experience from 'going out to work'.

Teleworking can, and does, have a more positive side, however, Margaret[12] was able to keep her job even though she relocated with her partner through negotiating a new flexible working arrangement with her employer. She spent a couple of days a week travelling to the sites she was responsible for, and the rest of the week she worked from home. She did comment that she received telephone calls from colleagues at 'all hours' but coped by switching on the answerphone at 7 o'clock in the evening!

Changing commuting patterns

In recent decades patterns of commuting have become increasingly complex. The spatial ties between residences and workplaces have greatly diminished by car-based mobility, as witnessed by the lengthening of journeys-to-work (Gillespie 1999: 27). Although most journeys-to-work are short, trip lengths are increasing in Europe and North America (in England and Wales for example, by 15 per cent from 1981–91), while the proportion of longer trips (over 30kms) increased substantially (by 45 per cent) (Bannister and Gallent 1998: 337; Green *et al.* 1999). In North America greater use is made of air travel for infrequent commutes (weekly or less regular commutes) than in the UK (see Chapter 2).

Among the dual career couples interviewed in the UK, two-thirds of the male partners in the 30 households had journeys-to-work which lasted in excess of 30 minutes, and of these one half had journeys exceeding one hour. Half the female partners reported taking at least 30 minutes to travel-to-work. Many respondents viewed an hour as an important 'psychological barrier': 'if it gets much beyond an hour you have to have a very good reason for it'.[13]

However three of the 60 British respondents had journeys of at least two hours in each direction. For one it was completely unacceptable. Unable to sell the family house to move closer to Nottingham (where he worked), he was now looking for another job. But for the other two such time spent commuting was acceptable; one said, 'I think it is part of the price of living out here' (a less accessible rural area). Clearly for many a long journey-to-work was a price worth paying for living in the kind of house they desired in their preferred residential location. Many consciously sought the peace and quiet of rural areas, as certainly happens in parts of Canada and the USA, such as in California.

Many interviewees had adapted to long journeys, some even finding time spent commuting as therapeutic. One woman who commuted 80 minutes to and from work by train on a daily basis used the train as an extension of the office: 'I've got well adjusted to it now ... for the work I do it's actually productive time'. The car is easily the predominant means of transport for travel-to-work in the UK, used by 24 of the 30 male partners and 28 of the 30 female partners. The flexibility, convenience and control offered by the car were perceived as its main advantage. Only a minority travelling to work by car actually needed a car during the day for work purposes. In the majority of households interviewed both partners had a car, and would find it difficult otherwise: 'we've had separate cars for years. I'd never manage without it now'.

Although there is still a marked pattern of commuting from peripheral areas to urban cores, along traditional lines, this is now paralleled by complex patterns of suburb to suburb movement, and even reverse commuting from urban homes to non-urban workplaces (Breheny 1999: 5; see also Chapter 4).

Living together apart

In Chapter 2 'commuter marriage' was examined as it produces a new social structure created because of the primacy of paid work. But in other dual career households living apart is a temporary phenomenon because of short term assignments (from a few days) some distance from the place of residence of the household. Dave,[14] for example, talked of one foreign assignment when their two children were young. 'I was asked to go out to Jordan and teach students who were qualified nurses in a big teaching hospital in Jordan. So I went there for ten weeks. I think that was probably quite a strain on Pam. She was very good about it. She didn't sort of put a step in the way. It was a very good experience but it was a long time for her'. Sometimes both partners have complex work patterns:

'I never stayed away unless I felt my wife was happy with me being away. Because of her job we had to balance commitments between the two of us. Even before the family came along, if she was working I would make sure I was around. We are very much a working partnership' (Ken).[15]

It is not only the labour market that is the trigger. The housing market can play a role in creating complex living arrangements, as appears to be the case for some workers employed in Silicon Valley, California. 'Declining quality of life in Silicon Valley – skyrocketing housing prices and crowded roads ... driving people out of the area ... are restructuring their lives ... the top echelons of tech workers may not live in Silicon Valley but they still work there' (Rabinovitz 1999). Those weekly air travel commuters from Austin, Texas to San José are called the 'nerd birds'. One large company employs staff in San José who divide their time between its offices in Palo Alto and homes in San Diego, Boston and London, UK.

One such commuter travels weekly from Portland, Oregon, an 80 minute, 675 mile journey every Monday and returns on a Thursday (*Sacramento Bee* 1999a: E1). He said, 'I can stay at work as late as I want. I'm not having to worry about getting home a certain time and tucking the kids in'. He can say this as his partner has become a solo parent midweek, and is now a full-time carer to support the dual location lifestyle. 'His' decision is based on more than a desire to maximise his spending power; 'I get to live a multi-million dollar Silicon Valley lifestyle that I could never afford in Silicon Valley'. The family home is a new 'five bedroom mini-mansion'. While he is 'home' he, 'can concentrate on the family, though sometimes 'work' intrudes on a weekend via e-mail. Midweek he lives in 'spartan quarters', where he just sleeps and spends most of his waking hours in work. His company does not contribute to the costs of his weekly commute.

Work and stress

Ginn and Sandell (1997) have reviewed studies of stress arising from the home-work interface (see also Lewis and Cooper 1983; 1988; Cooper *et al.* 1996; 1997). Such stress arises not only from pressures in the work and home environments but also from gender role attitudes internalised by individuals. Citing mainly studies from the USA on multiple roles they note that these are viewed as sources of either stress, satisfaction or both (Herman and Gyllstrom 1977). One approach emphasises the potential for stress of the combined demands of multiple roles, focusing on the adverse effects of role overload. Goode (1960) suggested that individuals attempt to avoid stress arising from conflicts between roles by abandoning or renegotiating them. Leaving paid work or working reduced hours would be examples of such a strategy, as was illustrated by two British women Jane[16] and Joanne[17] (for a fuller discussion see Chapter 2).

Other writers have argued that multiple roles may increase opportunities for social interaction and personal development (Sieber 1974). They claim that, in spite of some costs due to multiple roles, these are outweighed by the benefits. The conflict between the two perspectives, Ginn and Sandell (1997: 415) feel, is more apparent than real in that the crucial factor in creating stress is the total workload rather than the number of roles (Herman and Gyllstrom 1977). Thus the effect on women's health of combining work and home depends on weekly hours employed (Arber *et al.* 1987) and also on the way responsibility for the tasks of social reproduction in the household is defined and allocated (Wharton 1994).

In their study of stress amongst British social workers Ginn and Sandell (1997: 428) found that levels were related to 'home' (their family caring responsibilities) and to 'work' (paid employment). Having young children at home increased stress and was related to the age of the youngest child. To a lesser extent informal care-giving also raised stress levels. This source of stress is likely to increase with the ageing of the populations of advanced capitalist economies.

Recent research suggests that a rapid return to full-time employment after childbirth is confined to an élite of highly educated professional and managerial women (Ginn and Arber 1995; Chapter 2). These women are in households which are more likely to be able to commodify a number of the tasks of social reproduction (childcare, cleaning, etc.) (Mattingly 1999; Momsen 1999). On the face of it, their employment is little affected by motherhood, conforming closely to the pattern conventionally associated with men's employment. However, such women may pay a high price in terms of stress which, if sustained over a long period, is likely to entail risks to health. Among large employers in the USA there is greater recognition, than say in the United Kingdom, of the need for family-friendly employment practices in order to minimise stress and enable staff to carry out their jobs more effectively. Indeed employment law regarding the rights of women with dependent children and the provision of state-funded childcare varies between the three case study countries.[18] But both uncertainty as to when the working day will end and the sheer quantity of hours demanded are too much for some women as was illustrated by Joanne and Jane (for a full account see Chapter 2).

'Enterepreneurial' strategies of self employment for managerial and professional labour

'Entrepreneurial' workers are proliferating in the small business sector in all three countries, accounting for 16 per cent of the total labour force in Canada in 1998 (www.statcan.ca), 15 per cent of men and seven per cent of women in employment in the UK (ONS) and about eight per cent in the USA (Carnoy 2000: 75). As Sennett (1998: 118) comments, downsizing and re-engineering can impose sudden 'disasters' on managers and professionals so that a career can no longer be regarded as a 'well made road' (ibid. 120). Moreover the, 'short term flexible time of the new capitalism seems to preclude making a sustained narrative out of one's labours, and so a career' (ibid. 122). Sennett (1998) in his recent book described the collapse of the 'conventional career' through company-downsizing in up-state New York. A number of the dual career couples I have surveyed have also experienced downsizing at first hand. As was discussed in Chapter 3 James[19] had worked for three companies as a production director, and in each he had made people redundant. Indeed with the last two he was also made redundant. After being made redundant for the second time he became a self-employed business counsellor; for others telework opened up new job opportunities to be self employed.

Frank[20], a freelance editor, who lives in a rural part of the Midlands, UK, became selfemployed because his employer:

> 'relocated the operation back to London. I wasn't going to start commuting again, so I took the voluntary redundancy option and decided to go freelance. I'd been doing a certain amount of freelance anyway. So it was just a question of moving over to full freelance'.

He works in a profession with a long tradition of freelance work (Baines 1999). Technology and social networks have enabled some managers and professionals to become self-employed and take control of their lives (see Chapter 3).

Conclusion

Managers and professionals in Canada, the USA and the UK have a degree of autonomy as to where and when paid work tasks are undertaken; they are 'cash rich, but time poor' relative to other households. Those employees with control of their diary, i.e. a high degree of autonomy with time sovereignty, are content and less stressed. Younger employees, the so-called Generation X (Coupland 2000), are the group making the strongest demands for time sovereignty (Doyle and Reeves 2001: 17–18). This is especially so among young graduates whose numbers are increasing in all three countries. As undergraduates they tasted the time-sovereign life and are intolerant of the clocking-on culture of many employers (ibid.).

Achieving a balance between work and life has been the focus of a recent UK report by the Department for Education and Skills (DfES) (2000) and a measure of the UK Government's commitment has been the creation of the Work-Life Balance Team, as have some US universities. According to a British survey on work-life balance (Hogarth *et al.* 2000) greater flexibility of working hours is rated as more important to women returning to work than longer maternity leave. A range of policies introduced in the UK since 1997 focuses on women's growing independence and participation in the labour market, for example the five year National Childcare Strategy and extended maternity rights. 'Family friendly' employment policies are all too often equated with mother-friendly policies, however (Wheelock *et al.* 2000). In Scandinavia by contrast, there is paid parental leave, where a portion has to be taken by fathers and cannot be transferred to mothers, a more egalitarian way of facilitating the pursuit of two careers (ibid. Bunting 2000). Indeed women academics in Norway are entitled to sabbaticals every five years, and for male academics it is seven years.

6 Spatial mobility within the education system

Studying abroad: the globalising education sector

Bourdieu (1986) and others have noted the importance of the intertwining of cultural capital, social capital and economic capital with international migration (spatial mobility) in the construction of careers (social mobility) from/within the higher education system (see also Ackers 1998; Walters 2001 and Chapter 8). For those pursuing professional careers migration may well be undertaken to acquire extra qualifications at postgraduate level, as with those pursuing careers in medicine (for a fuller discussion see Chapter 8). According to Walters (2001: 7) the acquisition of particular symbolic capitals requires both an understanding of the possibility of strategic self-fashioning (Mitchell 1997) and knowledge of the global distribution of cultural wealth (cultural competence) (Hannerz 1996: 103). Proficiency in English and a 'western education' (attending the right schools and universities, acquiring educational qualifications) are assumed to represent 'the ultimate symbolic capital necessary for global mobility' (Ong 1999: 90 cited in Walters 2001: 7–8) (see Chapter 1). The focus of this chapter is on the ways in which international migration is undertaken to secure symbolic/cultural markers and this can occur before and after household formation.

Educational qualifications form the foundations of many managerial and professional careers today and are critical in shaping a person's ability to achieve upward social mobility. Moreover the acquisition of educational qualifications and striving to achieve upward social mobility can have an international dimension as some educational qualifications are 'portable' allowing access to labour markets abroad. This is the case for EU citizens within the EU migratory space as qualifications acquired from institutions within the EU have been harmonised (Ackers 1998; Hardill and MacDonald 2000). There is also considerable cachet attached to qualifications from prestigious US schools, or with the worldwide premium attached to information technology (IT) qualifications[1]. Some qualifications therefore are a means of improving life chances transnationally through spatial mobility (Ackers 1998; Beck 1992; Schoenberger 1997; Walters 2001).

Education can broaden horizons through travel overseas or what is known as 'OE' (overseas experience) to New Zealanders and Australians (Chant and McIlwaine 1998). Indeed OE has long been part of the educational experiences of Kiwi and Australian youth, prompted by the desire to learn about other countries (Pesman 1996). A central element of Kiwi and Australian culture, OE initially involved working in London, but now includes OE in a wide range of countries, including south and east Asia. One other facet of OE has been studying abroad, especially postgraduate study.[2]

In the post World War II period individual aspirations increased, as higher education became available to a wider number, especially in advanced capitalist economies. An OECD study (OECD 2000) has revealed that more young people than ever are graduating, with 35 per cent of each age cohort in the UK, 32 per cent in the USA and 24 per cent in Canada (ibid.).[3] In less than a decade in the UK the proportion of each age cohort graduating with a degree has risen from 20 per cent to 35 per cent (the equivalent in the USA was about a 2 per cent rise) (ibid.). Not only does education allow young people to seek more options in their careers but it also can open social and spatial horizons either as a temporary phenomenon or more permanently as migration may lead to, follow or precede household formation. Temporary international student mobility has been encouraged by a number of EU-funded schemes (such as Socrates and Leonardo, see also Chapter 8), and by the US Peace Corps.[4]

International spatial mobility for postgraduates is long established, as illustrated by flows of students migrating to the UK especially for higher degrees and/or specialist qualifications during the colonial era[5] or of flows of British graduates to the USA, Canada and elsewhere throughout the post-war period. But the international movement of postgraduates is escalating (and is now augmented by growing numbers of undergraduates and high school students spending a semester or so abroad). These flows remain predominately male (Bown 2000). While most students return to their country of origin, some overseas students build their careers from or in their country of study (depending on immigration rules for switching categories).

In 1980, 100,000 international students were enrolled in the UK; by 1996–7 the number had almost doubled. Today the international higher education market is estimated to be worth £1.5 billion to the UK economy (ibid. 9). In 1996–7, the USA had 458,000 overseas students (68 per cent of the global market), the UK 198,000 (17 per cent), Australia 63,000 (10 per cent) and Canada 34,000 (5 per cent) (Tysome 1999: 8). There is 'an insatiable demand for higher education, cost-effectively delivered, tertiary education in English' (ibid. 8).[6] The importance of English as a 'portable skill' is emphasised in a recent Ukcosa report (Ukcosa: the Council for International Education 2000) and by several interviewees.

One interviewee, Hela[7], migrated from Eastern Europe to the USA. Her education began in Eastern Europe, where she gained her first two degrees, but during her Master's degree through family contacts she spent a year living in the West (in the UK), 'It had to be an English speaking country. My French was good but my English was a bit better'. A second interviewee Tania,[8] an East European Junior Fulbright scholar emphasised the importance of a knowledge of English for career development. She applied for an American scholarship 'for me it wasn't difficult. I have knowledge of English. The older generation of professors don't [sic] know a foreign language. They were supposed to know Russian. The elder ones who retired knew French'. Qualifications are important but so is knowledge of English, or French in the Francophone world, as the two East European interviewees attest.

The USA's widespread global involvement after World War II increased people's awareness of the opportunities in the American higher education sector and indeed beyond (Manrique and Manrique 1999: 105). This was particularly true among the educated élite of Third World countries allied to the USA during the Cold War (ibid.). The National Science Foundation (NSF) estimated that in 1992 there were 49,000 foreign-born Asians and 3,100 foreign-born black Africans among the engineering and science PhDs in the USA. The equivalent combined number in 1962 was less than 1,600 (ibid.). Their knowledge of, and exposure to, American colleges and universities no doubt increased their interest in continuing study and professional development in the USA (ibid.). Furthermore political and economic upheaval has prompted some immigration from Africa and the Middle East for example.

Since the 1980s countries such as Australia, Canada and New Zealand have not only been important destinations for growing numbers of Asian immigrants, but the Asian economic miracle has seen the emergence of a wealthy Asian middle class (Ukcosa 2000). This middle class has become a purchaser of overseas secondary and higher education and English language training. Moreover they have been actively targeted by the higher education sector in the three countries (Pe-Pua *et al.* 1996).

Migration for the kids: astronaut families and parachute children

The British independent school system has long been internationalised, and has for many generations educated the children of wealthy families from the former colonies and dominions who boarded at these schools. This continues to be the case today, but is now augmented by children from countries with no historical ties to the UK, as well as by the children of skilled transients resident in the UK. These households value the ethos and education of these establishments, the delivery of education in English, and the acquisition of cultural capital imbued in such an education. In 1999–2000 there were estimated to be more than 15,000 foreign pupils studying at British independent schools (ISIS 2000). The independent school census of 1999–2000 recorded strong flows of new students from east Asia, especially Hong Kong, and from mainland Europe, especially Germany (ibid.).

The importance of the acquisition of the right symbolic/cultural markers has been identified as a key factor in the emergence of a transnational social morphology, including the astronaut family (Pe-Pua *et al*. 1996; Vertovec 1999). In the astronaut family it is not the male partner who migrates, but his family. He tends to return to his country of origin, where he continues his business or professional career. Astronaut families have been observed amongst members of the wealthy new middle class in Hong Kong and in other parts of Asia (Pe-Pua *et al*. 1996; Walters 2001). Emigration from Hong Kong to the USA, Canada and Australia has grown rapidly since 1980, and the astronaut phenomenon has been observed and studied in all three countries.

In an Australian study of 60 astronaut families (all skilled or business migrants), the majority had been to Australia previously and made informed decisions about their choice of destination (Pe-Pua *et al*. 1996). They arrived at a particular location in Australia where they had family or friends, and began settling into a new way of life – including purchasing a house, finding a suitable school for their children and possibly seeking employment. The overwhelming majority however returned to Hong Kong soon after arrival or within the first four months after arrival, and almost all did so with their spouse. The majority of the women subsequently returned to Australia to be with their children. A major reason for immigrants becoming 'astronauts' was the difficulty in finding employment appropriate to their level of training and experience. Failure to get overseas qualifications recognised, as well as being unable to secure and keep a satisfactory job, often led people to return to Hong Kong where job opportunities were better (ibid. xi).

In a more recent study of astronaut families in Canada, Walters (2001: 46) reported a different pattern to that observed in Australia in that the astronaut arrangement had been planned in advance of migration rather than reflecting a circumstance into which families had been forced after failed attempts to find work after migration. Overwhelmingly, the education of children was given as the primary reason for immigration, clearly implicating the strategy of cultural capital accumulation, in particular the education of children. Migration to Vancouver from Hong Kong and Taiwan, Walters (2001: 21) observes, does not reflect the traditional conception of migration for economic betterment; while the political instability of the home region was a major factor motivating emigration from Hong Kong and, more recently, from Taiwan, the security offered by a Canadian passport emerged in Walters' interviews, as did the accumulation of other forms of capital. Vancouver's 'environment' was cited, such as perceptions of 'safety', 'beautiful scenery' and a 'lifestyle' unobtainable in the country of origin, but of critical importance was 'education' (ibid.).

The parachute children attend private or state schools[9] as day students, especially where the International Baccalaureate is taught as in Canada and the UK. As was noted above, they live either with one parent (mother) or with a child-minder/nanny while their father, and in some cases their mother, live and work in another country. The household impact has been studied by Pe-Pua *et al.* (1996) and Walters (2001). These astronaut families were found to be different from other spatially separated family types, such as commuter, in that the family continued to operate as a unit (see Chapter 2). For women in astronaut families, the importance of having a network of friends and support groups was stressed. Walters (2001) found many examples of the intensification of traditional gender roles, such as the loss of the woman's economic independence (many had careers prior to migrating to Canada) and the increased personal undertaking of housework and child-care tasks. In some cases the move to Vancouver brought isolation, boredom, loneliness and strained marital relationships, with some male spouses keeping mistresses in Hong Kong and Taiwan (see also Yeoh and Willis 2000). For the children in astronaut families there is also a whole range of experiences, from shouldering more responsibility, to coping with differences in the education system, and missing a father figure, all of which sometimes lead to disruptive behaviour. The tendency of some astronaut parents to shower their children with gifts was also noted (Pe-Pua *et al.* 1996).

Janice and Nicholas[10] are the parents of three children, a son and two daughters. They are a West African dual career household she is an academic and he a government official. They were both educated to degree level in West Africa. Nicholas subsequently gained professional law qualifications in West Africa while Janice obtained a PhD in the UK (Oxford University). They have both have built careers in West Africa. Since Janice's student days at Oxford they have maintained strong research links with colleagues in the UK, and a London apartment, to which they travel regularly for vacations. When it came to choosing an education for their children, they opted for schools in the UK; the three children attend private boarding schools. They have the economic resources to contemplate such a choice which demands a large financial outlay. When the children have a vacation Janice travels to the UK and they spend time together in the UK the children come home to West Africa for the long summer vacation.

One of the key reasons Janice and Nicholas have chosen private schools for their children is that it will hopefully 'give them an "edge" in the UK job market'. Janice studied in the UK, her PhD is from an élite university (Oxford) and she feels she knows how 'things work' in the UK. The children attend an élite boarding school because she feels that 'they will have easier access to Oxbridge, by the time they are done you have been in the country for ten years and so are a resident and have no work permit issues'. They are thus hoping to give their children important cultural markers and social networks formed at élite education institutions in the UK as well as access to the UK job market without visa/work permit worries.

Solomon and Sandy[11] have five children and like Janice and Nicholas they were both educated to degree level in West Africa but travelled to the

USA for Solomon to study for his PhD. His career has kept them abroad they have lived in the USA, in Austria and the UK; for much of this time Sandy has been a 'trailing spouse'. None of the five children has been born in West Africa; one was born in the USA, two in the UK and two in Austria. While they have lived abroad she has focused on being a homemaker because of a combination of work permit and language barriers. As he noted, 'our kids don't know Africa, they have lived in the West nearly all their lives'. In 1992 he was posted back home to West Africa and they made the decision to keep their children in the British education system, 'to give them the best', and they thus live as an astronaut family. Solomon travels to the UK for Christmas, while Sandy and the children spend the long summer vacation in West Africa.

In another interview, in Canada, with a migrant professional couple from India, Ranji[12] emphasised that one of the key reasons for moving to Canada was non-economic, for the education (and future social mobility) of their children. Both Ranji and her husband Monder accepted downward occupational mobility as a 'trade-off' to give their children access to the education system of their host country; they have timed their migration so that their children will qualify for 'home student' fees when they eventually go to university in Canada. They anticipate their children will build their careers in the host country, as she said, ' there is no turning back'.

University study abroad

The ways is which education abroad has shaped the lives of postgraduate students is now examined through case studies to illustrate the complex motives for migration for education and training. Often very careful thought is given to both the choices of country and university for its 'cachet' and 'skill portability' and potential for building a career. The biggest international student movement occurs at postgraduate level, with the largest flows to the USA, but also significant flows occur to Canada, as my interviews attested (see also Tysome 1999).

Joanne[13] reported 'I came to [Canada] in 1994 to do a Masters. I was finishing at [university in UK] and I was interested in going away and I'd applied for a couple of places'. From making the decision to study overseas and securing a place with funding took several years. As to the reason why she wanted to study abroad Joanne said, 'I just knew I wanted to go overseas and I wanted the opportunity and experience to be overseas. Nothing appealed to me in Britain. I couldn't see the point of staying in Britain doing graduate work. I applied to Australia and to Canada. I didn't really know [much about the institution], I was just going on what other people said'. Thus for Joanne the desire for OE was very important.

The 'other people' was one key person, her UK dissertation tutor, who had contacts in Canada and Australia and he helped her: 'he wrote and spoke for me'. Social networks thus played an important part in shaping where she actually applied for postgraduate work. In fact the first offer she got was from a university in Australia, 'they offered me a chance to do my masters there but there was no money'. She put the decision 'on hold' for a while as she needed to secure a place with funding, and did some travelling first in Europe and then 'I came to Canada with Bob who was then my boyfriend. We travelled and explored North America'. While she was in Canada she actually used this travel time to check out departments: ' I came ... I visited and checked out the department. I was really attracted by the fact that there were [specific faculty members] so then I applied in 1993 and I didn't get accepted'. But with the help of department members she applied the following year and secured funding. Joanne migrated in 1994 and Bob joined her a year later. She has since gone on to study for a PhD and begin an academic career by securing her first lecturing post while Bob has established a small business.

One Japanese postgraduate student, the daughter of an academic dual career household[14] is studying at a US University because an American qualification will help her build a career in Japan, ' I want to teach at university in Japan and a PhD would be great from a US university. It looks good. So I came here. Also my major reason is I got offered a scholarship. I got money and opportunity'. The scholarship is a Fulbright. She applied, 'for three scholarships. I got all of them, Fulbright was better, more prestigious, better for my career'. Fulbright award 15 scholarships a year in Japan, and she got one of them.

She chose a particular university which she had already visited, 'I came here because the year before I came here for a conference and I presented a paper ... I met [faculty members] they seemed supportive of my project. I got some offers from other universities as well'. The USA was known to her as an undergraduate, ' when I was an undergraduate student I came to the States for a year. It was an exchange programme between my university and Wellesley College. It's a good school'. Her comment on Wellesley is rather an understatement, as it is one of the most prestigious women's colleges in the USA. What the year gave her was a taste of life: 'I had one year's experience of living here. Wellesley was a very competitive environment. I think I changed a lot. I became much more confident with myself. The professors [stressed] that you are going to be the leader [sic] as capable women. It was pretty inspiring'. She has thus lived and studied in the USA as an undergraduate and postgraduate, and through social networks that she has crafted she made the decision to study for a PhD at an East Coast university. The acceptance of the Fulbright scholarship was governed by what she thought would look best on her curriculum vitae (CV). In a very instrumental way at a very young age she is consciously building a CV.

Barry,[15] a British migrant to North America who left the UK in the 1970s to study for a PhD, has subsequently built a very successful academic career there. He, like many others (including me), had no clear plan for what he was going to do after his undergraduate studies: 'I wasn't sure and I applied for a number of different things and I wasn't really expecting to go to graduate school'. He applied for postgraduate work in the UK and North America (Canada and the USA). 'I made my final choice of going to ... by talking to one of [my UK] Professors. [Though he really had no knowledge about the department], he said it was a good place and thought I should go there. So I did'. He then reflected on the way he chose his postgraduate studies with the way students choose today, 'it was ridiculous when I think about it now and think of the way that our prospective graduate students come and scrutinise us. I had no idea about these different places I applied to. It was like choosing, a lucky dip as to where I ended up'.

International migration in academe

The immigrant in academe is not a new phenomenon. Immigrant professors in Canada, the USA and the UK have a long and distinguished record[16], such as Albert Einstein and the Nobel prize winners Chen Ning Yang and Tsung-Dao Lee (Manrique and Manrique 1999: 103). One particularly adaptive group of intellectual refugees fled to the UK after the 1956 uprising in Hungary, including George Radda, Chief Executive of the UK Medical Research Council and academic Imre Lakatos of the London School of Economics. Between 1979 and 1988 the number of Asian faculty in the USA increased by over 85 per cent, with most of the increase coming from foreign born faculty (ibid.). In 1993 alone, approximately 6,620 college professors were included among legal immigrants to the USA (ibid.).

The number of foreign-born faculty has also recently increased in the UK, doubling between 1994–7; 45 per cent of them came from continental Europe (Swain 1999). Other EU countries have not witnessed such an increase. The new faculty are largely employed at prestigious research universities, especially in clinical medicine, biosciences, business and management and nursing and paramedical sciences (ibid.). Such flows are thought to be the result of high job mobility in the UK created by the new research climate.[17] Short-term contracts are said to appeal to foreign academics at the start of their careers or when seeking promotion. By taking such posts in the UK they boost their careers when they return home.

Some migrants build careers after gaining a PhD in a host country while others migrate for career enhancement. As already noted Barry migrated from the UK for postgraduate study in the USA. During his first week he met fellow student Jenny whom he eventually married. 'I applied for a number of jobs in both in the USA and Canada'. They decided to stay in North America and 'yes I limited [the job search] to North America. In my dream world I would have liked to have gone back to Britain but first of all I don't think Jenny was especially interested in doing so. But there were no jobs. There was no way of me finding out about jobs. When I heard about the job market in 1983 it was really bad'. So a combination of personal and economic factors framed the decision: the fact that Jenny preferred North America plus the limited number of lecturing posts in the UK at the time he was looking for jobs. He also mentioned a third reason, social networks, 'I had [North American] contacts. People who write letters to you know people in the immediate vicinity. It's a network. So that is another reason why I stayed here'.

A second interviewee already noted also migrated from Europe to the USA. Hela's education began in Eastern Europe, where she gained her first two degrees. Through family contacts, during her Master's degree she spent a year living in the West (in the UK) when she returned to Eastern Europe, 'I went back ... after a year living in the West, and I could not settle. That did it. I tell you, once abroad, abroad for ever, I could not settle in'. But despite these feelings, ' I finished my Masters. I started to do my PhD but I had several lives. I was functioning, but I wasn't functioning. I was clearly looking for opportunities. I met somebody, who was in the United States, and I emigrated'. She emigrated during her postgraduate studies, married a fellow emigrant and resumed her studies in the USA. Hela completed her PhD in the USA but by then her marriage had ended and she tried to build a career in the USA but this was not easy, 'I still could not make the [assimilation] journey ... it took several years ... I was trying to assimilate by becoming an American [citizen]'. She had problems securing tenure because of her research topic, 'I didn't get tenure for the first time here'. But she secured tenure at the second attempt and has built a career at an East Coast university where she now directs a research centre.

Within academe international migration may occur for a short-term assignment as part of a sabbatical.[18] The sabbatical tradition within the academy is highly variable, however not all universities have such a system; some offer a one-year sabbatical but with an 80 per cent salary entitlement, others offer one semester at full salary. Academics from universities in New Zealand are actively encouraged to travel abroad during their sabbatical leave and receive travel grants from their institute. These monies are to facilitate international travel because of the geographic isolation of the country. A sabbatical can be very disruptive for the career of a partner and for the education of children, especially if it involves international travel. For these reasons, some people are very reluctant to travel far; some travel with the whole family and others travel as solos.

A sabbatical is rather like a short-term foreign assignment for someone who works for a multinational organisation. For a limited period of time an academic's skills and expertise enable them to become an 'invisible' skilled migrant, for whom national borders and barriers to the movement of labour do not apply. Sabbaticals involve travel for a finite period, and can cause physical and emotional disconnection from the work routine of their home place of employment. In addition, they work 'out of context' with limited teaching, administration; in my case I experienced virtually no disturbance from telephone calls and students.

One interviewee, an East European Fulbright scholar, Irene[19] was on a sabbatical at a US West Coast university when I interviewed her. For her the sabbatical, 'was a discovery, a new period in my life. To discover how to live in another country'. She found the whole period liberating after a very stressful period in her institute at home, 'I filled out the application ... I had been thinking about how to escape from [my work] for three years'. But the links with work are maintained by e-mail. She reflected on her first trip to the USA ten years ago when e-mail was not used as a means of communication by her, 'I remember ten years ago I just once a week wrote long letters [home]. I am pleased I am on e-mail. At the same time I need their letters'. Irene's mother is very ill. During the interview she talked at length about leaving her in Eastern Europe and her worries as to what she would do if anything were to happen to her.

Conclusion

In this chapter I have explored international migration among dual career couples for the acquisition of economic and cultural capital within the school and higher education sectors. The spatial mobility that economic and cultural capital brings has also been highlighted for those pursuing academic careers. While some qualifications are 'portable' others are not, and the lack of harmonisation of qualifications was emphasised by some interviewees. The EU is one of the few spaces where an attempt has been made to harmonise qualifications. The possession of language skills, especially English, is a second vital international 'portable' skill that facilitates spatial mobility for invisible migrants. Some households with the necessary economic resources send their offspring to be educated abroad (astronaut families and parachute children). For some today an international school education is acquired with the long-term aim of facilitating their children's entry to the host country's higher education sector and the pursuit of a career there without the worry of visa requirements.

7 Organisational careers
Patriarchy and expatriate work

Introduction

In this chapter I examine the impact on those dual career couples in skilled careers that require a high degree of international mobility, whether in the private sector through working for corporations or in the public sector (government and non-governmental organisations). A large number of professionals work abroad for international non-governmental organisations (NGOs), in the foreign service, such as embassies. In this chapter organisational careers and migration are examined with reference to careers within the military and for corporations; these moves are often gendered and have implications for the household. Contemporary corporate geographies have a strong international dimension. How mobile work is organised and managed, and how mobile workers are selected, has a great impact on mobile life (for a fuller discussion see Doyle and Nathan 2001). Multinational companies often require staff to be 'mobile', to be willing to work abroad, with international assignments forming an integral part of career development for potential senior managers. These assignments are relatively short term and are either followed by a return to the place of origin or transfer to another international location. These skilled transients are now a major feature of global migration systems, but they are mostly male (Adler 1994; 1997). Until the late 1980s, only five per cent of American expatriates, one per cent of Japanese expatriates and nine per cent of Finnish expatriates were women (Caligiuri and Tung 1999: 763). Two recent studies have revealed that the proportion of women expatriates is increasing. Florkowski and Fogel (1999) in a worldwide study undertaken in 1995 recorded that 11 per cent of respondents were women, while a more recent study of US expatriates recorded that 13.9 per cent were women (Caligiuri and Tung 1999: 764). Even though the proportion of women expatriates is increasing, it is still smaller than the proportion of women holding managerial and professional posts. Where career success is dependent on international assignments, women are therefore still at a disadvantage.

The expatriate management literature concentrates on 'the traditional expatriate 'cycle' – selection, training, relocation and adjustment, pay and performance, and return' (Brewster and Scullion 1997: 33). The process of adjustment has been described as a 'u' curve (Torbiorn 1982) with a 'honeymoon' period followed by a period of disillusionment. Studies have argued that the greater differences between the 'home' and 'host' cultures, the greater the likelihood of adjustment problems (Mendenhall and Oddou 1985). However, adjustment issues also arise among skilled migrants moving between European countries or to North America (Suutari and Brewster 1999). Though research on cultural adaptation has largely focused on employees in multinational enterprises and those pursuing organisational careers (Adler 1994; Forster 1994; Scullion 1994), skilled international migrants also include those pursuing occupational careers such as academics. Indeed coming to terms with new ways of working may have more serious implications for those skilled international migrants pursuing occupational careers than for expatriates with multi-national companies, who are working within the same organisational culture but in employment units in different countries (Hardill and MacDonald 2000).

As the international activities of organisations increase, so too do the expectations of organisations who send employees abroad (Darby 1995). A recent study of expatriate managers by Amrop International (1995) notes that most skilled transient managers are male, that 76 per cent are married, and that their female partners accompany them on overseas assignments. A study by Windham/National Foreign Trade Council (1995) reveals that 32 per cent of accompanying wives had careers before going abroad, however, only nine per cent managed to find work while their partner was on an overseas assignment. Those who did find work often took up part-time, unskilled or less-skilled work, to the detriment of their careers. When they returned home, their careers had often 'slipped back' in promotion terms, and they encountered problems in re-entering the labour market.

The adverse effect on the career of their partner may well be a contributory factor in the failure of managers to complete overseas assignments. Corporations cite 'spouse/partner issues' as 'the single most important hurdle in getting first choice candidates to accept assignments' (ibid.). Combined with 'family issues' these emerge as the most commonly given reason for managers leaving before the end of an assignment. The Amrop International (1995) study cited 'family concerns' as the greatest disincentive. These included the family's inability to adjust to the move, as well as the problem of finding employment for spouses (see also Darby 1995; Hardill 1998; Hardill and MacDonald 1998). International migration disrupts the second 'follower' career in dual career households and can compound the impact of the life course on the career of many female partners (Ackers 1998).

For many of the British and North American respondents who may be characterised as following occupational and/or organisational strategies spatial mobility and social mobility are strongly linked. There was some evidence from the interviews in the UK, however, that the need for inter-regional moves is weakening, 'there used to be days in the bank when if you wanted to get on you had to go to London. Those days are gone now'. Another bank employee suggested that the costs involved in the long distance movement of staff had become too prohibitive to continue with a large scale inter-regional movement. Despite this, over three-quarters of respondents from the British study indicated that they felt that career progression depended on their willingness to move, and increasingly that this could mean international moves.

When there are dependent children, decisions about whether to move or not may be even more complex. Many households underlined the disruptive influence on their children's education of making moves – particularly for children at secondary age. Moves may be delayed or turned down in order to provide children with continuity on examination courses. Moves may also be undertaken primarily for the education (and future social mobility) of children, as in professional householders who move from Asia to North America and the UK (see Chapters 6 and 8).

Dual career households and military careers

The 'ultimate' organisational career is one with the military. As the information displayed in Table 7.1 illustrates, the armed forces in all three countries have declined as a result of the 'peace dividend' that has followed the end of the Cold War. While the number of women in combat roles is increasing, they still comprise a small proportion of total numbers in the armed services in all three countries. For example, figures released for the British Armed Forces reveal that women made up 14 per cent of all recruits in 1997–8 (Ministry of Defence 1999). The armed forces in all three countries are not strongly feminised, and the position of gays and lesbians in the armed services remains contested. Pursuing careers in such masculine spaces is not easy (see p. 000).

The role of the armed forces is changing radically as a result of the 'peace dividend' (Lakhani 1994; Ministry of Defence 1998), but the pursuit of a military career remains basically the same: the service comes first and the household second. This view is illustrated by the manual *Wives' Guide to the Army* issued to the wives of British army personnel. The *Guide* (HMSO

Table 7.1 The armed forces of Canada, the USA and the UK, 1993–8

Military personnel for	1993	1995	1998
Canada		99,876	91,970
USA	1,716,432	1,533,116	1,403,114
UK	274,800	233,300	210,100

Source: Statistics Canada (www.statcan.CA/english/state/government/govt16a.html; http://web1/whs.osd.mil/mmid/mo1/smsir93.html; Annual Abstract of Statistics (1999), Table 4.3, 17

1994: 8) states that your partner 'is on duty 24 hours a day, 7 days a week … can be away from home at weekends, on public holidays, for your birthday, and for longer periods'. And that 'when you marry a soldier you must realise sooner or later you may be faced with the choice of keeping your own job or career, or 'following the flag' with him' (ibid. 57). It points out that 'moving is never much fun, but it is a part of Army life and has to be accepted' (ibid. 10). In no uncertain terms the wife of a British soldier is expected to be a 'trailing spouse' and to 'follow the flag'. But in return, the *Guide* notes that her partner has, 'a regular pay packet, and pay rates are reviewed every year. Longer holidays than most civilians; a place to live and people to ask for help and advice' (ibid. 8). Some of the highest divorce rates are recorded in military households (Hansard, 2 November 1999). This situation is now being recognised as a problem within the services and at long last the adverse effect of 'following the flag' is on the agenda (Ministry of Defence 1999).

Marchant and Medway (1987: 289) note that members of the US military are transferred from one military installation to another, often a great distance apart, every two to three years, as is also the case in the Canadian and British armed forces. They (ibid. 292) undertook an in-depth survey of 40 US army families, in which the male partner had had an average of 13 years service, moving an average of 10.7 times while in the military while their spouses had moved 7.6 times. Marchant and Medway (1987: 293) also found that for service members, the identification with the military way of life was a far more important predictor of 'well being' than was mobility history. They commented on some of the positive attributes of relocation with the military; not being faced with the 'stresses of finding suitable housing, schools, shops and services. Furthermore, military households are in a milieu where many people are new to the areas, where many are interested in meeting new families, and where the military provides opportunities for social gathering' (ibid.). While this is certainly true, the authors failed to acknowledge the adverse effect of severed social networks, school friendship circles, or consequences for the careers of partners of a new posting to another area/country.

Another US army personnel study undertaken in the early 1980s (McCubbin and Lavee 1986) of 1,000 families who had relocated from the USA to West Germany noted that there had been very little research by the US army on the topic of household relocation. It highlighted the weakness of US cultural adaptation programmes (at the time), along with the complexity and diversity of stresses and strains recorded by the surveyed families who experienced relocation at different stages of the lifecourse. The study recommended the adoption of cultural adaptation programmes that are sensitive to the life cycle.

A third study focused on the impact of the peace dividend on the US military (Lakhani 1994). The benefits of having more personnel based at home in the USA rather than in Europe and elsewhere were emphasised;

three of the five major positive conclusions focused on the military family/household. It was felt that 'home basing' (in the USA) would improve the quality of spouse employment as well as improving spouse and offspring quality of life (ibid. 124). While there may well be undoubted benefits of more postings in the USA, the problem of relocation every two or three years remains and will still impact negatively on the careers of partners.

One of the British dual career households I interviewed, James and Marie,[1] had tasted military life. James had a career with the British army when he and Marie began living together. Like most military careers, James's involved assignments abroad as well as separations caused by training and manoeuvres. These long periods of separation, as well as the partner's career often being put 'on hold', can put great strain on a relationship.

In describing their relationship together, Marie commented that in the early 1980s 'because of the army career we spent a lot of time apart'. After a year and a bit together, James was posted to Germany, and Marie's career suffered. In reply to the question, 'what was happening to your career', Marie replied, 'nothing, that was it. I had to give it up, I was a housewife. It was frightful because I had no identity of my own. It is an identity crisis for many women'. Added to this there were separations, 'when we were there I hardly saw him. He was always on exercises. The army was the main problem ... it's a dreadful thing. It does terrible things to families. It is very hard ... it's almost normal to break up. We were conscious that we weren't getting enough out of our relationship ... so he left the army'.

James acknowledged the problems for Marie in Germany: ' Marie is a pharmacist, so worked wherever we went, other than when we were in Germany where it was impossible to get any work, what is legal in one country may not be legal in another, so she still wouldn't be able to practise in Germany without taking a top-up diploma or certificate'. The career problems caused by lack of harmonisation of qualifications, which affect many international migrants, were compounded by army life. James went on to say, ' it was a bit of disaster being in Germany. A combination of alienation, two young children and me on a job that took me out of the house all day. We really did not want to go overseas again after that'. As a result James left the army.

The separations that go with life in the armed forces were described by others including in literature produced for British families in the armed forces, 'my husband has joined the 'Bosnia bunch' [UN duties] for six months and I am balancing the needs of three young children, job, dog, car, money and my sanity' (*Army Families Journal* summer 1998: 1). This woman's story highlights the many ways in which her partner's posting impacts on the family unit. With the father way for six months, she is responsible for holding the family together and for all the tasks of social reproduction as well as holding down a job. These issues have been highlighted as a cause for concern in the

British Government's Strategic Defence Review initiated by the Labour administration of Tony Blair (Ministry of Defence 1998; 1999).

Some couples adapt to 'commuter marriage', like newly weds Dan and Kerry[2] (see Chapter 2). They are unusual in that they both have trained as fighter pilots with the Canadian Air Force. Kerry, 'was a trail blazer', one of the first women to train as a fighter pilot. She began training after her first degree at the age of 21: 'at the time I wasn't interested in a long-term relationship. I was more interested in flying fighters'. She talked about flying fighters, 'when you climb into the cockpit of an F18 you don't need strength. You need endurance; hand–eye co-ordination'. Despite this, the air force and the world of fighter pilots in particular remains very masculine.

She left after her first tour because 'I was getting exhausted, being the only woman' but she also had other plans, 'of becoming a doctor and becoming involved in medicine'. Some of her colleagues were envious of this move, both of her willingness and ability to make it: 'some of my colleagues [pilots] were married with children and couldn't afford to take the dangerous step of loss of income. The people that did protest interestingly were the wives of my neighbours [pilots].'

She met Dan, 'a month after I got out of the military' when she had begun studying again some distance from Dan's base: 'he had dreams and aspirations of flying fighters as I did'. Since they became a couple they have never lived in the same town but have sustained a relationship. They meet up 'every month; sometimes we have gone as long as six weeks. We try and alternate it'. She talked about how they communicate: 'we use the phone, phone each other every day; he doesn't have easy access to a computer'. As they live some distance apart they have to consider time zones for contacting each other, 'we are only an hour apart this time of the year. [But] it is awkward, we sort of know each other's routine'.

Kerry has concerns about the military way of life, 'about what the military environment does to people. I think it is very isolating, contrary to what people might think', but she also has a deep understanding of the life of a fighter pilot, having been one herself. They married in spring 2000 after maintaining a 'long distance' relationship for over five years. They know that because they prioritise their two careers – fighter pilot and cancer researcher – they will continue to live as a commuter couple because air force bases and suitable research centres for Kerry to work in are not located in the same places.

Living abroad: the impact on the household

Living arrangements

Working abroad is very variable experience. It occurs in diverse settings (Findlay and Gould 1989) and moves take place for an array of reasons

(Hardill 1998). Most women migrate as tied migrants, as illustrated by Adler's (1994: 29) survey of 686 American multinationals. Further, 'over half the companies reported that they hesitated to send women managers abroad ... and 70 per cent believed dual career issues were insurmountable' (ibid. 30).

In some locations such as the Middle East foreign workers live in compounds, in a 'bubble' cocooned from the world (see p.98). Life there has some parallels with life (for Westerners) in the colonial era and diplomatic life today. Life elsewhere for foreign managers and professionals can be within the 'host' communities, in which some migrant household members are left feeling very isolated and alone because of linguistic and cultural barriers. This can occur even when their partners' are still using English and have the same cultural mores in their work for their multi-national company regardless of location. Adjustment problems for the household can cause some households to return 'home' early.

When offered a global assignment, dual career couples must decide for themselves the conditions under which they would accept. The couples, especially the trailing spouse, often consider a number of options:

- turning down the global assignment, although this often results in negative career consequences for the partner;
- finding the trailing spouse a position in the foreign country. Ideally both partners move abroad at the same time, although more commonly they move at different times, with the timing of each move based on the separate needs of each career, or for wider household reasons (such as educational needs of children);
- finding the trailing spouse a position in the same region of the world, thus becoming a commuting couple;
- having the trailing spouse remain at home in his/her current position, thus becoming a commuting couple;
- having the trailing spouse take a sabbatical, a career break that allows him/her to accompany his/her partner;
- creating other options that work for the particular household (Adler 1997: 280–1).

Migrant households therefore adopt a variety of living arrangements. Some move *en masse*, while others opt to live at dual locations as a 'commuter couple' because of the transient nature of the work. Such an assignment is normally for a few months and because of difficult living conditions and/or in the interests of the career of the partner and/or education and welfare of the children, the two partners live as a commuter couple. International moves can be disruptive to family life, especially for women and children who have to adjust to life in different cultural settings and with different levels of emotional and physical support. Friendships and the security they bring to the young are disrupted by these moves. Living in dual locations as a commuter couple imposes emotional stress and strains, including problems within relationships (Yeoh and Willis 2000), but does allow the spouse to maintain a career. These issues are now explored along with the impact of skilled international migration on the careers of the two partners.

Hans and Bettina[3] have spent the last four years living on the West Coast of the USA. They have always prioritised his career. Bettina has always been a 'trailing spouse' having withdrawn from the labour market when the children were born. She used this period as a personal opportunity to retrain. She was studying at a German university when Hans' job offer in the USA appeared, 'it was a good career move for him'. When asked how she reacted to the idea of moving to the USA, 'it was mixed … I thought a lot about the kids [then aged 12 and 8 years] and I was one year short of finishing my Master's in Germany'. 'I had to make the decision if I would go with him or not. He was pretty much determined to go. I don't know if he would have gone without me, we never discussed it at this point. It was a big decision to make and I knew that my daughter was totally against it, totally opposed it so I had to do it against her wish'. Although Hans was fluent in English and Bettina's English was also good, their children could literally not speak a word.

In saying yes to the move Bettina put her husband's wishes before those of her daughter and indeed her own personal aspirations to obtain a new qualification. But she did make certain conditions. 'One of my conditions, I said I would only go if he stayed with us until the end and we moved together. I don't want him to leave [me alone] with the kids. Then we came together and had two weeks looking for a house'. Bettina was very insistent that she had a say in choosing their place of residence. There were a handful of other German families working at the West Coast plant, and Bettina said that 'two [of the migrant women] had big problems. They moved to … and it's only a village with nice houses, swimming pool, but no library, no coffee shop. It is typical suburbs. The funny thing is with these two women the men came before without the women and they chose the house. I said, I don't want that. I want to choose the house I'm living in'. She chose a house in a town with a network of facilities, where other German women had advised her to live. The town has a university with an international centre which she used to develop a circle of friends.

Hans' employer gave them, 'one week cultural training and they had workshops on [the USA] and on the move. They paid for the move and they organised it'. While, 'the inter-cultural sessions were very helpful' nothing was targeted at the children; it was up to the parents to deal with that. One of the problems Bettina had was dealing with her daughter's hatred of the USA before and after the move, 'She was severely mad for at least three quarters of the year. She was just not like herself. Not willing to accept things that Americans did. She tried to be as German as possible. We were not allowed to speak English. We had to speak German. She was not willing to let me help her. She was crying the whole time'. Her daughter was devastated to leave her friends in Germany. Ironically, now it is her daughter who does not want to return to Germany. But the whole episode meant that for the first year, 'my daughter took all my energy'. Then Bettina was able to organise her education, a US Master's degree.

When I interviewed Bettina (April 2000) she was facing another trans-Atlantic relocation, this time back home to Germany: 'we could stay longer [in the USA] but ... that would mean my husband has to be employed as an American employee'. But then there would be financial penalties, 'health insurance in Germany, our pensions, it's a financial decision. He is guaranteed a job, at a certain position, a certain level, managerial level. That's not the case if we stay here and go back on our own later'. They were due to return to Germany during the summer 2000. At the time of the interview in April 2000 they were unsure what job Hans would be offered and where exactly they would be living.

The refusal of work abroad

In one of the British dual career households the male partner had been asked to relocate abroad by his company but refused following lengthy negotiations. Jane and Mark[4] are an egalitarian household. He has worked in the UK for a multi-national, 'an export job for a company who was owned by a Japanese company'. They faced a very difficult period recently, 'I had been doing the job for about a year ... they wanted to restructure ... which meant the area I was looking after was going to be handled from Amsterdam, The Netherlands'. They asked me to relocate to The Netherlands immediately and I said, 'yes we will do that if the money was OK'. According to Mark, his boss, 'wanted to put me up in a hotel and pay for everything'; but said 'oh no' when Mark asked about paid flights home at a weekend: 'I wouldn't have seen Jane for three months. But that was perfectly acceptable to the Japanese'.

Mark's account illustrates that differences in cultural expectations regarding acceptable terms and conditions of relocation between his Japanese employer and himself. When asked if Jane would have been able to get a comparable job in the Netherlands (as a food technologist) Mark said: 'no I don't think so. She may have fallen lucky, but of course the language is a problem there. I would have struggled, but being a Japanese company, the company language is English'. In the end he decided not to relocate, and it was 'a tremendous relief all around'.

Thus for Mark and Jane, a dual career couple, the offer of work in the Netherlands was not accepted because Mark was not satisfied with the terms and conditions. His career prospects with the Japanese company were obviously blighted by his refusal to move to the Netherlands and he was forced to find alternative employment. Interestingly, Jane did not mention this potential move. Instead she focused on her current problems of balancing a career with childcare issues.

The 'expatriate' bubble

One important location for skilled transients is the Middle East. Two of the case study households have lived in 'expatriate compounds' in the Middle East. Some households adjust to life in the 'bubble'; others find it too oppressive. Bill[5] is a Quantity Surveyor who accepted a posting with an international construction company and 'spent 8 to 9 years overseas ... we retained [our] house [in the UK] ... because working overseas, a lot of people think it is really glamorous, but you work hard as well, one day you can just be out of work'. They spent the first two years in dual locations, then Betty joined Bill in the Middle East where their son was born. She could not obtain a work permit so that when she moved to the Middle East to join Bill she had to withdraw from the labour market. Betty enjoyed expatriate life and she became a full-time carer in the Middle East, 'although Saudi Arabia is a huge place it is very easy ... for shopping, you don't have to queue'. Betty was not allowed to drive a car in Saudi Arabia and Bill went on to say, 'a typical day she'd get up, walk the dog, game of bridge, that's it'. Jenny and Peter[6] have also lived the 'expatriate life' with Peter's posting to the Middle East. Jenny commented that, 'you call it a camp, but in fact its just like a housing estate ... the husbands work for ... a lot of domestic help, whatever help you wanted you got'. She went on to say, 'some people hate [it] it's so claustrophobic but I loved it ... swimming pool, restaurant ... [an] absolutely fantastic life [for] children ... I wouldn't mind going back tomorrow'.

Both households recounted memories of life in the 'expatriate bubble', of a world within a world, very reminiscent of the expatriate life of the colonial era (as portrayed in the writings of Paul Scott 1966, for example). While they worked in the Middle East their lives centred on the closed area of the foreign (Western) compound.[7] Indeed when describing their life with the multi-national company, Jenny noted that moving from country to country means that, 'you are having to change values all the time ... [when] you're a real expatriate ... you all stick together like glue and lots of English is spoken ... [but] when you are not really in the expatriate bubble you are trying to live in a community like the Netherlands you have lost your identity and you are amongst Dutch people and you don't fit in in quite the same way. Not all migrants adapt to expatriate life, especially when their careers are put 'on hold': 'my first problem was getting a work permit ... [plus] I couldn't drive ... very bored as I'd always been used to working and being independent'.

The commuter couple

For some household members work locations abroad are very difficult to cope with. For Jenny in The Netherlands: 'the house design ... the urban life and I think it was the big change ... flat, wet miserable ... so much to cope with ... in a normal community [no] bubble anymore but you are in a foreign place and they are talking around you in a foreign language. And I think I was so worn out with the children'. The loss of identity and security provided by expatriate life in the Middle East, combined with emotional and linguistic barriers proved too much for Jenny in the Netherlands. This also combined with problems with house design, which was typically Dutch: 'a huge downstairs room ... vertical stairs, when you have young children it's not a very good place to live'.

She went on to say that on the first day she moved into the house her middle son broke his arm falling downstairs. Issues of education compounded these problems. Peter's company offers to pay for the education of his children at boarding schools in their 'home' country. But Jenny did not want this, 'ordinary folk like me it's not part of our tradition ... it's very hard to send them off to boarding school especially so young, lots of people do it, they are entrapped in their [expatriate] lifestyle'. Peter's employers offer very different fringe benefits to Mark's. Peter's company covers school fees and weekly commuting costs. Moreover, they recognise that the emotional wellbeing of spouse and family is a shared responsibility between employer and employee. Not all employers have such policies.

Following Peter's job relocation to the Netherlands, Jenny became very unhappy. She was effectively a single parent, as Peter's job demanded a heavy investment in time. She made the decision to return to the UK, with Peter's full approval, and they now live as a commuter couple. Their 'family home' is a large modern detached house in a small market town in the East Midlands, UK, very close to her parents who help her almost daily with the care of the children, ' he works in The Netherlands, comes back on a Friday and goes back on a Monday ... we have four or five airports within striking distance'. She went on to say that 'since we have been married we have been all over the place ... it gets to the stage where nowhere feels like home and you feel very much subject to what the company wants to do all the time'.

Fortunately Peter's employers recognise the importance of ensuring that their workers' families are happy; they pay for his travel costs to the UK every weekend and allow him to arrange his working hours to accommodate his weekly commuting pattern. Peter's weekly commute involves a car journey to a British airport, a flight to Schipol and then a train and tram journey to the office every Monday morning and the return journey on a Friday afternoon. During the week he lives in a small flat close to the office, where he works very long hours three days a week. This 'dual location' lifestyle means that he compartmentalises 'home' and 'work' (Green *et al.* 1999). But he does use travel time – the air travel element – as office time, and occasionally works at home on a Sunday, mainly reading reports and preparing for the week ahead. Jenny still has sole responsibility for the tasks of social reproduction. The help with childcare since returning to the UK has enabled her to resume work in careers guidance. She currently has a part-time job which she enjoys very much. While Peter and Jenny are a long-term commuter couple, a number of the surveyed households in both

North America and the UK indicated that short-term assignments abroad had made them temporary commuter couples (see Chapter 2).

The return home

A number of returning skilled transients described in other studies (Salt 1988: 392) mentioned 'reverse culture shock' and other problems of adaptation. The return home featured strongly in my interviews with former skilled transient households. Susan[8] for example, who returned to the UK in 1993 with her husband Rahul, said 'we tried to make a joint decision, but I am sure that in many relationships there is always a little bit more of one person in any decision than another'. Toward the end of their spell abroad she worked as an educator in nursing, a position she found, 'challenging ... I established a department and developed educational programmes'. She went on to say, 'if you are coming back into the country it is very difficult to get a job at the same level you left. Or at the same level that you'd been functioning at in the country abroad. We both made the decision [to come back]. We both made endless lists, reasons for staying, reasons for going. Financially coming back here has been a struggle'. While Rahul had found a job before they returned Susan, 'didn't know that I would have a job'. She took responsibility for organising the move back to the UK. When they returned she was out of work for seven months, and then found work in the Education Department with the neo-natal service. She felt that she might well have progressed further in her career had she not gone abroad. She works full-time and needs help with the boys and so employs a nanny, whom she is responsible for hiring and paying (Momsen 1999). House hunting upon their return to the UK centred on villages within the catchment area of a good state school (as indicated by their friends) for their sons.

With Bill and Betty their choice of home – at some considerable distance from Bill's place of work – was also very much the result of the catchment area of good state schools. The return home has seen lifestyle adjustments for both of them. Betty looked back nostalgically to life in the Middle East, 'when we came back to the Midlands ... I used to hate it ... queuing in the shops ... queuing to park the car'. Bill has attempted to improve the quality of his life by becoming self employed and controlling the amount of travelling he does, while Betty now has a part-time job which she combines with domestic chores. She finds it very rewarding. Theirs is very much a 'traditional' household, with Betty being prime carer, partly because of Bill's need to travel for his job, but also because Betty derives fulfilment, 'quite happy just doing part-time work. There's too much to do really. A big garden to do and the house needs cleaning. If I went full-time I don't think I would cope. So I think part-time is just about my barrow'.

Conclusion

While human resource departments look for ways of making overseas assignments acceptable to dual career households, the economic pressures on global companies may pull in the opposite direction. Couples who are already abroad may find that repatriation means redundancy and may be reluctant to put themselves in that position. At a conference for expatriate women in Brussels (FOCUS 1996), many women living and working in Europe and beyond felt that their partners had no choice but to accept the conditions offered by their companies. This was especially so when they were no longer in their thirties or when they worked for companies who were 'downsizing' in their home country. Many female partners felt that the possibility of negotiating for the maintenance of their careers was not a reality in the present economic climate. Yet the projected number of skilled transient managers seems set to increase. Foreign assignments appear to be emerging as a vital part of career development for both senior managers and professionals, and women often appear to be denied them.

A recent study on expatriate adjustment and commitment has focused on the ways host country employees and expatriates interact (Florkowski and Fogel 1999). It highlighted an often overlooked aspect of expatriation the commonplace but contentious practice whereby multi-national companies extend much more generous reward packages to expatriates than are offered to their functional local counterparts (ibid. 785). Strong perceptions of pay inequity and rivalry over career advancement opportunities may also explain why some high status locals perceive resident expatriates more negatively than lower status host country employees (ibid.).

When it comes to selecting staff for international work, most companies focus on the individual, looking in particular at their skills, ambition, leadership potential and career development (Doyle and Nathan 2001: 40). Study after study highlights intercultural adaptability as one of the key skills for international workers and present family factors as one of the main reasons that overseas assignments fail (ibid.). Scullion (1994: 100) points to another problem, namely the increasing concern that multinational companies express about 'growing reluctance to international mobility'. In a study of 45 companies (based in the UK and Ireland) he found that 'concern about dual career problems and disruptions to children's education were seen as major barriers to international mobility'.

Accountant Peter Morgan (1989: 94–5), in a highly personal account, describes his relocation from London to Sydney, Australia by his employer, noting that working abroad can adversely affect both careers. An increasing number of companies have a written or unwritten dual career policy (about 40 per cent in a survey by PriceWaterhouseCoopers) stating what assistance they will offer an accompanying partner (cited in Doyle and Nathan 2001: 41). Mobile work is very intense however, partly because it tends to involve tight deadlines and can compound the stresses of a 'hardworking' organisational culture. The costs of expatriation are high. The average costs of relocating an employee have been calculated as $US 60,000 (Windham/National Foreign Trade Council 1995: 22). With such large sums at stake it is important that companies choose staff who are going to stay the course. In addition to cross-cultural adaptation programmes (Darby 1995), other strategies focus on selection and involving partners at all stages of the discussion. This is seen as increasingly useful as employers come to recognise problems arising from dual career partnerships. But it should be remembered that it is not just relocation abroad that imposes career compromises. Recent research in the USA among professional workers (Malecki and Bradbury 1992) found that organisations are reporting increased resistance to relocation (within the USA) amongst dual career households. The subject of labour mobility, in general, therefore, is problematic for dual career households.

8 Professional careers and skilled international migration
Case studies of healthcare professionals

Introduction

This chapter focuses on the members of two professions, medicine and nursing, in order to examine households in which men and women are pursuing occupational careers. While the status, rewards and prestige of doctors and nurses are different, for some considerable time members of both professions have migrated internationally in an attempt to secure social mobility (Commonwealth Secretariat 2001: 1). Sir William Osler (cited in Verney 1957) has stressed the merits of overseas travel for the career development of doctors. Popular culture representations such as the US television series *ER* also visibly portray overseas-qualified doctors (from the UK and former Yugoslavia) in the emergency room in the Chicago public hospital, along with US-trained ethnic minority doctors. There are differences in the mobility of doctors and nurses, however. Doctors tend to move for longer periods of time and over greater geographical distances than nurses, and their mobility is more likely to be emigration in the strictest sense of the word. Nurses tend to move shorter distances,[1] make less permanent moves and remit more of their earnings home (Commonwealth Secretariat 2001: 1).

While universally medicine is judged as an élite profession and top doctors are well paid, nursing is not accorded such status. As was noted earlier when Americans are asked to rank the most desirable jobs, consistently they place doctors just below Supreme Court justices, the top-ranked (Sullivan 1995: 2). There is broad agreement on the prestige of doctors in other countries including the UK and Canada (Susser and Watson 1975). The status of nursing does vary from country to country because of differences in their role and level of education and training. A survey of the Asia-Pacific region (Johnson and Bowman 1997: 205) found that while most countries classified nursing as a profession, the terms paraprofessional and associate professional were used in Australia, Indonesia and Singapore.[2] This chapter takes up the nature of professional careers in the health fields and explores the role of overseas recruitment within them.

Professional careers, gender and professional regulation

Medical care today is highly complex and costly. This is due in part to technological change, to the ageing of the population in all three countries and the associated high call on medical services, as well as the rising expectations of the population as a whole regarding what the health service can/should do (Sawada 1997). The health services in all three countries are highly organised, with bureaucratic administration. Core state funding through national health insurance supports provision in Canada and the UK. In Canada, health provision is a provincial responsibility, with some variation in levels of remuneration (Figure 8.1). In the UK, a private health sector exists alongside the state-funded National Health Service (NHS). No private sector exists in Canada. In others countries such as the USA, the health service is largely financed by self provisioning through individual and employer insurance schemes, but there has been some state funding for elderly and low-income populations through Medicare and Medicaid since 1966. Whichever system of funding, the health service today in advanced capitalist economies needs more and more financial support.

While nursing has long been one of the most feminised occupations in most countries, the medical profession is now feminising, linked to the opening up of higher education to women (see Chapter 6). In the UK, for example, women comprise 29 per cent of doctors and half those currently training in medical schools (Crompton and Harris 1998: 302). Universally, access to medical school is academically highly competitive and medical training is arduous, taking between six and seven years to become fully qualified, with a medical degree and professional registration. Specialist training takes even longer; in the UK, for example, the level of Consultant or Principal in General Practice, is not usually achieved until a doctor's mid thirties. In the UK students enter medical school straight from high school in their late teens, while in Canada and the USA medical students are older, in their early twenties, as medicine is a postgraduate degree. In the USA students usually follow a science course before medical school.

Medicine has been described as a 'vocation, or calling' by Sir William Osler (Verney 1957: 4; as has nursing, see below). Medical careers demand commitment, and often involve working long hours, especially during the post-registration specialist training period which in all three countries involves studying and working at the same time after acquiring a medical degree. In a recent report on the relation between work and family life among 202 UK hospital consultants a number of single women highlighted that they had had to make choices between their relationships and their career during their training years. One commented that 'in earlier years I made decisions about my relationship because of my job and training ... having to choose between a relationship and a job at that time' (single female consultant, childless, surgical speciality). Another talked about the difficult choice: 'this was the relationship

Figure 8.1 Crisis in nursing – press cutting from the *Vancouver Sun*, 28 June 2001
Source: © United Nurses of Alberta

where settling down and family were important. He needed to move to
Cardiff [Wales], there was no job for me ... this was the nightmare you
never wanted to be in "is my career more important than my relationship?" For me I had a strong commitment to medicine and I was at the

stage when I wanted a consultant post' (single female consultant, child-less, medical speciality) (Dumelow *et al.* 2000).

This study also noted that frequent geographical moves were a key factor in hospital training at the time the consultants were training. Currently in the UK the use of rotational training schemes has reduced the frequency of geographical moves for doctors who are in post-registration training. The surveyed male consultants were more likely to have made frequent moves for their career than female consultants (half of male consultants compared to one third of female) reflecting the fact that a greater proportion of male consultants than female had 'trailing spouse' partners (see also Hardill and MacDonald 1998). As one indicated, 'in the early days my wife suffered. I recognise a strong debt to my wife and to the children' (male consultant, psychiatry speciality). In contrast, female consultants were likely to remain in a geographical area with access to a range of potential jobs so that both partners could pursue careers (ibid.). As one said, 'I made a conscious deci-sion to stay in London, recognising that we had to find two medical jobs and there would be more in London' (female doctor, dual doctor partner-ship, diagnostic speciality).

Even though the profession is feminising, it remains highly segregated by sex in all three countries. Male doctors predominate in the most prestigious and highly paid specialities such as surgery. Women doctors tend to be clus-tered in 'feminine' or 'family-friendly' specialities such as paediatrics, dermatology or radiology which offer 'convenient' or at least regular hours of work (Crompton and Harris 1998: 302). Crompton and Harris also argue that the long period of training required of doctors has had the ten-dency to be associated with 'forward planning' of the domestic career of women doctors (ibid. 306) who opt for flexible and part-time work (see Chapters 2 and 5). Doctors have traditionally been expected to plan their professional and domestic lives in line with Osler's view of a medical career, 'heavy as are your responsibilities to those nearest and dearest, they are out-weighed by the responsibilities to yourself, the profession and to the public' (Verney 1957, cited in Dumelow *et al.* 2000: 1,437). Indeed Osler advo-cated that medical students not form relationships while studying advising them to 'put your affections in cold storage for a few years' (cited in Verney 1957: 24).

The UK hospital consultants study also included an examination of the factors that consultants themselves considered important in shaping the bal-ance between their work and their personal or family life (Dumelow *et al.* 2000). About one fifth of women but only three men believed that their family or personal life had been restricted during their twenties and thirties, either intentionally or unintentionally, for the benefit of their career and that they had a career-dominated relation (ibid. 1,439). Some expressed strong regrets at giving so much time to their career. One said, 'I wouldn't start my medical career again. It's taken too much out of me' (single, female, childless, surgery), while another said 'I'm aware I gave too much to medicine and lost out as result' (divorced, female, childless, obstetrics and gynaecology) (ibid. 1,440).

Two-thirds of the men and just half of the women had highly segregated professional and family lives. Family responsibilities were organised to prioritise careers. This involved a spouse's offering full-time domestic support or the use of paid domestic help (ibid. 1,439). Many men in this group, however, were dissatisfied with the balance between their career and family life; as one said: 'most clinicians in acute specialities work ridiculous hours, at the expense of their health and their families. But if you want to pursue a career in acute specialities that's the way it is' (married, male, children, surgery) (ibid. 1,440). About one tenth of the men and a third of the women had an accommodating relation between work and family. All the women in this group had worked part-time or taken a career break at some point in their career. The men had restricted their work commitments or limited their career goals to benefit their personal or family life. Most men and women in this group expressed satisfaction with the balance between their career and family life (ibid. 1,439). As one said: 'I didn't stay at work any longer than I had to. I was meant to go out for drinks after work with others including senior medical staff, I was meant to socialise with the team, but I didn't. As soon as my shift was over, I was off home to my family. My family were important' (married, male, children, diagnostics). This quotation highlights the importance of social networking and the blurring of work and leisure time involved in building a medical career.[3]

Stereotyped opinions and prejudices about women in medicine are prevalent amongst senior members of the profession. Informal networks that influence the appointments process persist (Allen 1994). Their influence is assisted by the dominance of medical appointments committees by internal senior staff. Though Allen (1988) was writing about the reasons for the continuing low representation of women in senior levels of the medical profession in the UK, the situation is similar for overseas-qualified doctors, especially ethnic minority doctors (Anwar and Ali 1987).

In order to obtain professional registration doctors, whether they have qualified at 'home' or overseas, have to pass an examination of the regulating medical association. Medical school graduates first obtain limited registration to cover a period of specialist training which is 'on the job' and follows graduation. Upon completion of specialist training they then apply for full registration and in some cases membership of the college of their specialism (such as surgery or general practice). Only those whose qualifications are recognised by the host country can begin the process of obtaining professional recognition, and this process can involve examinations, further training and considerable outlays of expenditure (see Table 8.1, p.108).

Nursing has long been regarded as being one of the 'caring' professions[4] (Poole and Isaacs 1997). It has also been seen as an 'appropriate' career for women, as an extension of the kinds of duties that women performed in the home as wives and mothers, incorporating both love and labour (ibid. 536).

Table 8.1 Professional regulations in medicine

Country and regulatory body	'Home' doctor registration rules	'Overseas' doctor registration rules
Canada and the Medical Council of Canada (MCC).	Following completion of a medical degree doctors must pass the Qualifying Examination Part 1 to assess their competence for supervised clinical practice in postgraduate training programmes. Criteria for eligibility: • Graduate at a medical school accredited by the Committee on Accreditation of Canadian Medical Schools • Graduate from a US medical school accredited by the Liaison Committee of Medical Education (LCME)	Doctors must: • Get permission to immigrate to Canada • Sit Evaluating Examination of the Medical Council of Canada • May have to undertake between 2 and 6 years medical training in Canada • Apply for a license, and pass part 1 Qualifying Examination • Following 12 months of post-graduate training sit part II Qualifying Examination
UK and the General Medical Council Medical graduates obtain provisional registration for the first year of post graduate training before applying for full registration.	Doctors entitled to GMC registration if they: • Qualify at UK medical school • Qualify elsewhere in the EU	Doctors can apply for limited registration and must: • Pass the Profession and Linguistics Assessment Board (PLAB test) along with an English (IELTS) test • Have a job offer • Obtain a place on a GMC approved training scheme • Provide evidence of achievement in specialist training
USA and State Medical Boards (National Board of Medical Examiners).	Graduates of US, Puerto Rican and Canadian medical schools must pass the United States Medical Licensing Examination (USMLE). This is a common evaluation system for all applicants for medical licensure and is required for the initial license to practise medicine.	Educational Commission for Foreign Medical Graduates (ECFMG) certification assesses the readiness of graduates from foreign medical schools to enter US residency and fellowship programmes The following criteria must be met for certification: • Pass examinations in medical and clinical science • Pass English language proficiency examination • Pass Clinical Skills Assessment (CSA) • Provide documentation of medical education at a medical school listed in the World Directory of Medical Schools

Source: http: //www.gmc-uk.org; http: //www.mcc.ca: http: //www.ecfmg.org; www.nbme.org

A clear career path still does not exist for most nurses, and so nursing is still said to offer a 'discontinuous career' (Bonney and Love 1991). In Japan the average nurse has only 3.9 years initial continuous service (Sawada 1997: 247). This lack of a clear career path lies at the heart of the current labour supply problems in nursing in many countries, including Japan and the UK (Hardill and MacDonald 2000; Sawada 1997). While most nurses enter the profession not 'for the money' but because of a desire to nurse, these feelings in an increasing number of cases fail to sustain a career (Hardill and MacDonald 2000; Johnson and Bowman 1997; Sawada 1997). Nurses are often frustrated by low pay, lack of available resources, inappropriate workloads and inefficient support systems (Sawada 1997; Walter 1989). While some leave the profession, others use their 'portable' skills to work abroad. In Canada it is estimated that almost 50 per cent of nursing graduates work in the USA, and about one third of those who stay in Canada leave nursing within three years of qualifying (Bailey and Culbert 2001).

A number of writers have claimed that historically nursing has been a subordinated occupation, a semi-profession differentiated from the 'full' professions by virtue of shorter training, lower status, less 'privileged' communication with clients, and a less developed body of specialised knowledge (Carpenter 1993; Etzioni 1969). Skills rather than knowledge are stressed within nursing. Nurses are highly managed and controlled within their working environments 'the fact is that nursing is part of the medical hierarchy, and however far you rise, you can't get full responsibility: the doctor always knows best' (Coward 1997: 15). Johnson and Bowman (1997: 203) conclude that while many professional attributes have been attained in nursing, two unresolved issues are developing a unique knowledge base and autonomy in practice.

While nursing is seen as a semi-profession, in common with medicine, clear standards affirm professional status. Professionalism is thus seen as a form of occupational control. Regulatory bodies deal with professional standards, offer advice and consider allegations of misconduct. The maintenance of a professional register is a central feature of the regulatory system in the three countries (see Table 8.2, p.110).

The international migration of doctors and nurses

There is a long-established tradition of doctors and nurses being skilled transients. Hezekiah (cited in Dvorak and Waymack 1991: 121) notes that from 1890 until 1940, medical and nursing services were developed by a branch of the British Civil Service, whereby doctors and nurses were recruited from the UK to serve in the then colonies and dominions. In the post-war period, highly complex global flows of qualified and partially-qualified doctors and nurses as well as of student doctors and nurses have developed. Concerns about shortages of medical staff and the tremendous drain on resources that can occur when skilled health professionals migrate,

Table 8.2 Professional regulation in nursing

Country and regulatory body	'Home' nurse registration rules	'Overseas' nurse registration rules
Canada; College of Nurses of the province e.g. Ontario. Registration is for life.	Nurses who have completed an approved nursing course in Canada or the USA have to apply to the College of Nurses for the state they intend to work in and then pass a licensure examination.	They must have: • Completed a nursing programme comparable to one in the province; • Provide evidence of recent practice; • Pass registration examination; • Demonstrate fluency in English or French; • Show proof of Canadian citizenship, landed immigrant grant status or authority under the Immigration Act (Canada) to practice nursing.
UK and United Kingdom Central Council (UKCC). Registration is updated every 3 years for UK and EEA nurses and every 2 years for other nurses.	Nurses who have completed an approved nursing course apply to UKCC for registration. No further examination is necessary. Provide a certificate of good conduct.	EEA nationals (plus Iceland and Norway) with EEA qualifications can apply for UKCC registration and no further examination is necessary but a certificate of good conduct is required. Nurses from USA, Canada, New Zealand, Australia, South Africa, The West Indies and Hong Kong may be immediately accepted for registration providing certain criteria are met. In addition they will require work permits/visas. All other applications are considered on an individual basis but the candidate may be required to; • Gain further clinical experience, either in UK or abroad • Undergo a further period of education and training • Pass an English test
USA and National Council of State Boards of Nursing.	Nurses must apply for licensure to the state or territory in which they wish to practice and satisfy the board's eligibility requirements prior to taking the National Council Licensure Examination (NCLEX). NCLEX test centers administer the examination, which tests the knowledge, skills and abilities to nurse. This examination is administered in USA, American Samoa, the District of Columbia, Guam, the Northern Mariana Islands, Puerto Rico and the Virgin Islands.	To be eligible for the licensure examination overseas nurses must have proof of: • licensure/registration and being in good standing in another country • or official documentary evidence of successful completion of a nursing programme

Source: http: //www.cno.org/; http: //www.ukcc.org.uk; www.ncsbn.org

especially from the Third World, was recognised at the Commonwealth Medical Conference held in Edinburgh in 1965 (Commonwealth Secretariat 2001: 1). Subsequently international disquiet about this 'brain drain' led to the establishment of the WHO Multinational Study of International Migration of Physicians and Nurses, the findings of which were published in 1979 (ibid.).

To varying degrees, the health sector in a number of countries is sustained by overseas-qualified[5] staff. At one extreme are hospitals in the Middle East, where staff from up to 40 countries may be working in one hospital (www.arabiancareers.com), receiving levels of remuneration linked to their last country of practice. Some doctors and nurses migrate to countries like the USA or the UK for further training to augment qualifications acquired in their country of origin. These new qualifications and experience can enhance career progression back in their country of origin, in the host country or indeed elsewhere, especially in the Middle East. As most doctors and nurses migrate in their twenties and thirties, migration can lead to, follow or precede household formation. Further, the formation of a household in the host country can influence future migration patterns (Anwar and Ali 1997; Findlay *et al.* 1994). In addition, some doctors and nurses migrate not as 'migrant professionals' but as tied migrants ('trailing spouses') or for household re-unification and therefore appear in other immigration categories (Hardill and MacDonald 1998).

The international migration of doctors and nurses is heavily regulated both by state immigration rules and by professional regulation (see above).[6] State immigration procedures are detailed on websites for health service workers, for example www.homeoffice.gov.uk/ind/dad.htm for doctors and dentists wishing to migrate to the UK; application forms can be downloaded (www.homeoffice.gov.uk/ind/afm/htm). Moreover country specific websites are maintained, for example for US-qualified staff detailing employment opportunities in the UK (www.britain-info.org/bis/fsheets/7.htm).

Major political shifts have influenced the territoriality of labour markets in the post-war period, the most striking example being that of the EU. Only within the EU have cross-national attempts been made to standardise the education of doctors and nurses (see Tables 8.1 and 8.2, pp.108, 110). Article 8 of the Treaty of Rome states that 'every person holding the nationality of a Member State shall be a Citizen of the Union', and shall, as a consequence, 'enjoy the rights conferred by this Treaty' (paragraph 2 cited in Ackers 1998: 90). The close link between Community citizenship and mobility is reflected in Article 8a which outlines the fundamental right 'to move and reside freely within the territory of the Member States' (cited in Ackers 1998: 91). During the early phase in the development of the European Economic Community (EEC), the free movement of workers was viewed primarily in economic terms, as a means of removing barriers to labour mobility and to provide the same treatment across the EEC in matters of employment as to nationals of each State (ibid. 94).

The mutual recognition of qualifications has therefore provided a framework for the movement of healthcare professionals – doctors, dentists, pharmacists, midwives and general nurses – between EU states since the 1970s (Seccombe *et al.* 1993: 125). At the same time, the nature of the relationship between EU countries and their former colonies and dominions, some of which were suppliers of doctors and nurses, and other destinations for qualified doctors and nurses, has been redefined. Interestingly, Canada has re-aligned its medical registration procedures to the US system.

While state rules and professional regulation are important, recent work by Tyner (2000) on Filipina nurses has highlighted the importance of a complex network of participants who facilitate international migration. This network includes government officials, private recruiters, foreign employers and migrant workers themselves. The process of recruitment involves a number of stages, namely contract procurement, labour recruitment and worker deployment. Some specialist agencies exist for nurses (Tyner 1999) and/or for specific destinations (such as Japan, Kuwait or Taiwan). Tyner's (2000) work also illustrated the importance of social networks which act as conduits of information. They can be global with the use of the Internet, for example the websites of alumni associations[7] or involve intensive use of e-mail (interview material, see also Grewal 1996).

Recruitment agencies also play a key brokering role in matching the supply and demand of doctors and nurses (Tyner 2000). Agencies focus on specific geographical areas, or on the supply of one national group (Tyner 2000; Figure 8.2). While doctors for the Middle East are recruited through

---

The Middle East
Arabian Careers Ltd (www.arabiancareers.com) specialise in recruiting doctors and nurses for Saudi Arabia.

England
Health Professionals is a London-based medical recruitment agency. They place overseas nurses seeking work in UK hospitals.
http: //www.healthprofessionals.com/

New Zealand
Wheeler Campbell Consulting Ltd is a broadly-based human resources and recruitment company placing locums and permanent health professionals, via international and internal recruitment, throughout New Zealand.
www.healthrecruitment.com/

Europe and the Middle East
Kate Cowhig International Recruitment of Dublin works for client hospitals in Europe (including UK) and the Middle East recruiting nurses, doctors and other medical specialists (radiographers, medical secretaries etc.). http: //www.kcr.ie

USA
Professional Placement Resources, based in Florida, actively recruits from English-speaking countries for USA hospitals. http: //www.pprjobs.com

---

Figure 8.2 Medical recruitment agencies

agencies, for other locations vacancies appear in journal advertisements and there is an open competition, except for locum posts which are often filled by agency staff. Agencies are important in the recruitment of nurses. The remainder of this chapter examines the nature of these diverse migration flows drawing on published research as well as individual migration stories. I use the term 'overseas' doctors and nurses to mean those who were born outside the country in which they are practising/living and who have migrated to that country either to study or to obtain a medical or nursing job.

Overseas doctors and 'portable' skills

The international migration of doctors has a relatively long history predating recent globalisation processes. Overseas-qualified doctors have complex migration trajectories. One study found that the international migration of Hong Kong doctors was conditioned by differences in visa and settlement policies and professional entry conditions in the three major receiving countries of Australia, Canada and the UK (Findlay *et al.* 1994). India generates a massive surplus of doctors, the result of demand for medical education that is part of a general demand for higher education, unrelated to employment opportunities. In 1980 in response to the growing unemployment of graduate doctors, the government of India decided not to increase the number or capacity of medical schools. This decision resulted in the rapid growth of private medical colleges that now constitute 17 per cent of all medical colleges in India sustaining the surplus production (Commonwealth Secretariat 2001: 23). The major exporting schools however, are the élite public sector institutions with high international standing and grant support, such as Baroda University Medical College, the Seth S.G. Medical College in Bombay, the Madras University Christian Medical College in Vellore and the All India Institute of Medical Sciences in New Delhi. In terms of scale of migration from these institutions, it has been found that 50 per cent of graduates from Baroda University Medical College spent their most productive years in the USA and at one time Bombay graduates constituted 11 per cent of all Indian medical graduates in the USA (ibid. 23). The salaries they can command in the West are four to five times what they earn in India (Oommen 1989: 416–17; see also Parekh 1994).

Research from the USA reveals that overseas-qualified doctors form a significant segment of highly-skilled migrants entering the USA, with the largest group of overseas-qualified doctors coming from India (Kanjanapan 1995). These flows can cause disquiet as Richard Sennett (1998: 127) notes 'the fear that foreigners undermine the efforts of hardworking native Americans is a deep rooted one. The global economy serves the function of arousing this ancient fear. Many American physicians have cited, for example, the flood of 'cheap doctors' from Third World countries as one of the reasons their own security can be menaced by insurers and health-maintenance companies' (ibid. 127).

Migration of doctors into the USA is not only from Third World countries however. Grant and Oertel's (1997) study of the supply of Canadian doctors highlights an absolute shortage, especially in rural areas, which has been aggravated by the migration of Canadian doctors to the USA. These flows are long established, and were facilitated by the relaxation in American licensing and immigration laws after 1962. Flows have increased again recently because of the rapid expansion of Health Maintenance Organisations (HMOs);[8] which are being used in the insurance schemes offered by a growing number of employers. Tracking the 'class of 1989' Ryten et al. (1998) found that seven years later 11 per cent of the 1,722 graduates of Canadian medical schools were practicing outside the country. Emigration was particularly acute amongst specialist doctors.

A consistent feature of healthcare policy in Canada over the past 25 years has been a resistance to using overseas doctors (Grant and Oertel 1997: 163). Accordingly, the number of immigrants arriving in Canada with the 'intended occupation' of doctor fell from 1,300 in 1969 to under 530 after 1976, or less than one per cent of the stock of Canadian doctors (ibid. 164). Overseas doctors have been almost exclusively used to address under-servicing in specific regions and specialities (there are some parallels with the situation in the UK). Newfoundland, Manitoba and Saskatchewan have historically relied on overseas doctors to offset the loss of doctors to other provinces and countries. In 1996, there were more overseas doctors practising in Saskatchewan than there were domestically-trained doctors. Hiebert's (1999) analysis of migration flows to Montreal, Vancouver and Toronto however highlighted the importance of 'corporate élite and overseas doctors and lawyers'. A strong cluster of Vietnamese doctors can be found in Montreal (for linguistic reasons) and of Indian, Chinese and Jewish doctors in the other two cities.

In Australia, prior to 1966 immigration policies technically excluded people on the basis of language not race, but intent and practice discriminated against Asians, and the policy was highly effective in maintaining a 'White Australia'. Gradual concessions were made after 1956 and then 1966 when applications were permitted from an array of professionals, with further relaxation in 1972 (Monk 1983). National need was soon defined to include professionals and business people. In her analysis of Asian migrants, Monk found that, occupationally, the non-European groups consisted chiefly of overseas doctors, other health professionals and engineers, scientists and academics. Most doctors settled in the cities of Sydney, Melbourne and Perth, though doctors practised in country towns. Their settlement patterns are thus very similar to those observed for Canada.

Monk (1983: 6–7) also analysed the residential pattern for those in Sydney, and found them to be dispersed throughout the city. Given that they are professionals, the degree of dispersal across socio-economic status areas is interesting, and is the result of many practising medicine in low-status neighbourhoods while the native-born serve more affluent populations. The immigrants have encountered difficulties in gaining full recognition of their professional qualifications and experiences unless these were gained in the UK, the USA or a few approved Asian universities (ibid. 14).

Reliance on overseas doctors is much greater in Britain than Canada and Australia. This situation has developed because the supply of doctors from British Medical Schools is less than the number of doctors required by the state-funded NHS, and has been so for several decades.[9] The shortage has been exacerbated largely by limitations on the working hours of junior hospital doctors as well as the long-established flows of UK-qualified doctors abroad. The key destinations are the USA, Australia and the Middle East where salaries are more lucrative and conditions of employment better (Hardill and MacDonald 1998).[10] Data from the General Medical Council (GMC)[11] show the number of registered doctors in the UK increased by 48 per cent from 1986 to 1995, but that the proportion who qualified in the UK has fallen from 61 per cent to 42 per cent (Fletcher 1997: 1,278). In recent decades, however, a number of changes in British immigration rules and the accreditation of qualifications have altered the source countries from which overseas doctors originate (Iredale 1997; see Table 8.1, p.108). These processes are leading to a complex but little documented shift in the intake of doctors within the NHS (Jinks *et al.* 1998) which is reflected in the numbers of doctors registering with the GMC. Between 1986 and 1995, the number of UK-qualified doctors only increased by two per cent, while the number of European Economic Area (EEA)-qualified[12] doctors trebled and doctors who qualified elsewhere doubled (Fletcher 1997: 1,278). The scale of the desire to obtain Western qualifications results in a 'blurring' of education and work and is illustrated by the fact that 1,500–2,000 overseas doctors enter the UK each year in order to undertake post-registration medical training and examinations.

Monk's (1983) study of Indian migrant doctors to Australia illustrates the complexity of the migration motives. She found four basic important recurring motives for migration to Australia. Push factors associated with political conditions at home (East African Asians in the 1970s) were the prime inducement for one third to migrate, and also entered into the decisions of others. Some anticipated problems for their children. For these migration meant giving up established professional practices, separation from relatives, and changes in domestic and social life, particularly if the family had been used to domestic servants. Nevertheless, most seemed to have a realistic view of their situations, to be satisfied with their choice, and saw their move to Australia as permanent (ibid. 10). Thus non-economic factors, namely wider household wellbeing, were prioritised before the medical career in migration motives (for a fuller discussion see Chapter 6).

Those who moved from India had been subject to less pressure to migrate but dissatisfaction with circumstances at home, particularly with career prospects, played an important part in migration decisions and provided a second recurring motivation. Some were members of minorities – Parsis or Christians, for example – who considered their 'backgrounds' to limit opportunities. Others felt restricted by limited research facilities, low salaries and not having the 'right' connections (social networks). Competition for good positions is severe in India but these people thought they had the qualifications for professional success in an appropriate context.

Attractions in Australia provided a third motivation. Pursuit of higher education and specialist training was a major motive for some though a number of them had already studied and worked in Britain, Europe or Canada. While they seemed likely to settle in Australia, they also anticipated opportunities for further international travel to maintain links beyond Australia. The final motive, for a relatively small number, were their personal ties with Australia. These included those coming for marriage to Indians, those with Australian spouses, or people joining parents or children who had already migrated. All these intended to stay.

My own interviews with doctors similarly reveal complex histories of migration, training and employment. Sameera, Ali and Judy did their pre-qualification training in one country and migrated to the USA or the UK initially for post-qualification work. After obtaining the qualifications Sameera[13] returned to Pakistan to pursue her career, reaching the post of Assistant Professor of Paediatrics at a Medical School. Her partner Mushtaq also began his studies in Pakistan where he obtained a science degree and then went to the USA to study for a PhD. In 1991, he moved to the UK after obtaining the post of research fellow in plant molecular biology. He has a full-time fixed term contract at a Midlands university. It was when he had secured this post that Sameera came to the UK in 1994 to marry him. She has been able to obtain a work permit and has a job as a Clinical Assistant working 14 hours a week.

Sameera thus moved back to the UK for personal reasons, to marry Mushtaq. International migration on the part of Mushtaq and Sameera appears to have adversely affected her career as the 'follower' (Bruegel 1996; Hardill *et al.* 1997a; Hardill and MacDonald 1998). Although she held an Assistant Professor's post in Pakistan, her resignation from that post was within the context of the longer-term possibilities of intertwining her career development with household formation in the UK. Although she is within commuting distance of a number of hospitals, the only post she has been able to get is part-time and fixed term as a Clinical Assistant looking after children with cancer. She is currently undertaking research in an effort to try and improve her job prospects.

Ali (from Iran) met Suzanne[14] in Scotland in 1982 when he, 'was post-qualification and she was pre-qualification. I was slightly in front'. They married in 1988, but Suzanne said, 'we've only been living in the same place since 1991'. Ali said that he 'didn't train (pre-qualification) here, I had my training abroad [in Iran] and had a job before [in Iran]' coming to the UK. He obtained a series of qualifications in the UK, including membership of the relevant Royal College. When he came 'I was on a lower grade than [what I had been on] abroad'. Thus migration for him, like Sameera, was initially followed by downward occupational mobility.

Comparing his post-qualification work experience with that of his wife Ali reported that while Suzanne held a series of six-month posts in hospitals in lowland Scotland his 'job has been so much more continuous. I never changed from the same unit ... short-term contracts as well ... but in the same place'. He went on to say that, 'Suzanne's career is normal amongst members of the medical profession [six month contracts in different hospitals demanding spatial mobility] because of my background I was probably more senior than what I was doing so they kept me on because of that'. Thus Ali's experience was recognised (but not financially rewarded) by his team and for that reason he was offered a succession of six month contracts with the same team. This is not unusual for overseas doctors.

Suzanne, 'chose neurology ... and followed the established career path ... but the big bottleneck is going from Registrar to Senior Registrar ... there are only two places in Scotland which have big neuro set-ups'. We then talked about gender and specialisms in medicine. Suzanne said that 'in neuro-surgery at the time [of planning her career] there was only one female [consultant] neuro-surgeon in the UK ... [now] I think there are four. It is a very strong male preserve'. She went on to say, 'If I'd been male I would probably have persevered and struggled on to clinical neurology ... but as it was having decided to get married and I wanted a family [and got a consultancy as a neurophysiologist which] is very competitive, and there is a long wait for consultant posts ... I was almost nine years waiting'. Suzanne, like many women doctors, knew that surgery was largely a male preserve, and chose within her area of interest medical neurology rather than neuro-surgery, seeing this as a realistic branch in which a consultancy was achievable by a woman. We should remember that many men and women overseas doctors do not make it to consultant (Anwar and Ali, 1987; see also Sameera's comments on p.116).

Suzanne has had three children (when she held Senior Registrar and Consultant posts) and has taken three months maternity leave for each child. Each time 'they organised the funding and I found the person [to act as locum]... I took a month's annual leave, and I worked right up until the last days. It was really the last day'. When the first child was born they were a commuter couple, 'Ali in Edinburgh and then Birmingham ... I had my babysitter in Edinburgh, my job in Glasgow'. But she feels 'as far as the career goes, the children haven't interrupted my career at all'. Ali's parents planned visits to the UK to help with the arrival of each grandchild, 'for the first three or four months I was back [after my maternity leave]'. Suzanne is prime carer, and Ali contributes very little (time or money) to the tasks of social reproduction.

Ali's marriage to Suzanne, a British citizen, suggests that it is very unlikely that he will leave the UK. Similarly Mario's marriage to Joanne[15] has led him to settle in the UK, but as will be seen (p.119), he also had career progression problems in Italy. The impact of marriage on the mobility of migrants was also noted by Findlay *et al.* (1994). They found that married migrants were likely to have worked abroad for longer periods than those who were not married, suggesting that marriage creates ties which reduce a migrant's mobility (ibid. 1,616). Suzanne and Ali, like a growing number of dual career households, have found that managing two careers may demand spells as a commuter couple, as well as long-distance commuting (see Chapter 2).

Career blockages at home featured in a number of migration stories. The blockages centred on issues of gender but also financial opportunities. Judy, the daughter of dual career couple Sam and Victoria,[16] is very conscious of her Nigerian identity, but is also very much part of a global family. Since 1994 her father has worked in the USA and her mother in Nigeria, 'they shuttle back and forth. Both my parents are Nigerian, but in the 60s [sic] they were able to get a Green Card at that time. I was born in upstate New York'. She went to Medical School in Nigeria, 'in a class of 120, we had about 50 women'. But her studies were marred by political unrest. 'The year before I left there were riots ... changing oil prices or something. Schools were closed down for five months. A lot of my friends went to the UK or America to work, just to do odd jobs, just to make up the time'. Judy used her social networks to enable her to travel and to find temporary work. She stayed with a long-standing Nigerian friend in London. Her temporary move was facilitated by her dual nationality. 'I was in London, I worked as a nursing assistant for three months. I got it through an agency. I had a hard time because I was not British'.

She returned to Nigeria to complete her medical degree, then did a compulsory internship in the country. It was during this period of hospital work that she and her class mates built on their recent experience abroad and their social networks 'we talked about what we were going to do next year [we] started finding out about examinations'. She has very negative feelings about the social constraints placed on single women in Nigeria. 'We didn't want to go back home to live. The options were limited. Back home [Nigeria] it's not very easy, a single girl on their own is not looked upon very nicely in Nigeria. Then the stories of visa problems started coming back. Everyone was holding on to whatever they had. [For me it was the fact that] I was born in New York'. She discussed with her parents the merits of going to the USA. 'When I was doing my internship I discussed with my father [about moving to the USA]. But he thought of it as an opportunity, so in December 1991 I came to the USA'.

With the financial support of her parents she prepared to sit the US medical exams while living with family friends. Her Nigerian university qualification was recognised, 'they have a list of countries with universities qualified to do the exam' so this entitled her to prepare for the US professional registration examination. 'They used to have the National Board Exams, which even their medical students take. They now have the United

States Medical Licensing Exams. It's the same exam for everybody. I was [in] the first set to take that exam' (see Table 8.1, p.108). While she was preparing for the exam she also obtained paid work 'as a medical assistant at a health centre. There were a lot of African doctors there. So we were able to interact a lot, they gave me a lot of pointers'. Those pointers were so important as 'they have different regulations needle stick protection, stuff like that we didn't have in Nigeria'.

When she passed the examination she was then able 'to apply for a residency – a speciality, I applied for obstetrics and gynaecology and paediatrics. It's very hard to get into obstetrics and gynaecology, a lot of spots are saved for American graduates, it's very high paid, very stressful, [it is important] to have a reputation from here [USA] all the time. I applied for 80 obstetrics and gynaecology spots all over the place and 20 paediatrics. Eventually I was interviewed at a hospital in Long Island [for a paediatrics residency]. They had housing there, right in the hospital grounds'. She thus accepted a residency in her second preferred speciality and began building a career in paediatrics and to develop strong feelings of attachment to the USA. During her residency other members of her family moved to the USA for work and study which 'helped a lot'. During one visit home, 'I realised Nigeria wasn't home anymore. My mother was there, but it wasn't home anymore. I actually wanted to come back [to the USA]'.

Judy has used newly acquired social networks to help her build her career in the USA, 'while I was in residency, my chief resident, she's not Nigerian, but Indian, she helped me get my Fellowship Training in Emergency Paediatrics' [in New York]. Judy and her Indian friend both now work in a hospital in New York, 'I work in the Emergency Room, it makes you able to handle a lot of emergencies'. But life is not plain sailing; 'there is a lot of racism still. Some patients who are African-American have a hard time dealing with an African or African-American doctor. They have the idea that you don't know as much as a white person'. But issues of gender are also a cause of anguish too, 'I have a name tag that tells you what my role is, I had a day when I worked with a male nurse, he got called the doctor and I got called the nurse. It was fine you know. In the city there are a lot of people trying to make changes. I think that's my role, in many ways, to make sure I change things'. Of her 120 medical school classmates, 'maybe about 20 are in the USA and there are up to 50 in the UK'. The fact that their Nigerian medical qualifications are recognised in both the USA and the UK has enabled half the class, those with the economic resources and contacts, to migrate and to build a career outside Nigeria.

Mario, like many of Judy's class mates, now practises medicine in the UK. He first came to the UK in 1991: 'I had recently graduated and started my career in ophthalmology in Bari, Italy. And there were some grants available from divisional authorities to go abroad. The idea was to go somewhere in the United States and do something specific about retinal diseases. But the Director ... didn't have any contacts in the US, and they had a very weak contact through another resident with ...[hospital] in London'. He went to the hospital: 'actually it was the best thing that had happened in my life. I really enjoyed it. I was supposed to do four months, but I stayed for seven and [since then I have] kept going back to continue the research project I had started'.

Mario was obliged to return to Italy in 1992 to complete his military service, after which he resumed his career in Italy but he, commented, 'when I went back I felt I was different from everybody else I had developed a different mentality'. This new perspective was reinforced by his response to the Italian medical regulatory environment 'largely dominated by nepotism. So I came back here [to the UK] ... which happened by chance. I was in Rome and a Consultant I had met at ... [UK hospital] rang me to say would I be interested in being a locum for six weeks. It was August or September, too hot in Italy, not much was happening, most hospitals are closed apart from emergencies. So I thought okay, I can have a mixture of holidays and work'.

Mario's initial entry to the UK, like Sameera's and Ali's, was to obtain extra qualifications, to develop his skills in his chosen research area. He never anticipated building there, and indeed, returned to Italy. He does however feel that the environment in the UK for a medical career is superior to that in Italy. His return to the UK was largely thanks to his informal contact network. Since the first locum post in 1992, he has held a whole series of posts as a locum, research fellow, Senior House Officer, Registrar and Senior Registrar in hospitals in five English cities (London, Bristol, York (where he met his wife), Derby and now Nottingham). He and Joanne now live in the Midlands, UK and are expecting their first child.

Mario explained that he is now consciously building his career in the UK: 'I can use my specialism diploma [obtained in Italy] to be recognised on the specialist register here [in the UK]. That means I can apply for Consultant posts. But I felt I needed more surgical training. After checking with the President of the Surgical Specialist Committee, I realised that I could compete with anybody else for more surgical training. I am now seeking to actually get British qualifications ... to make me feel more part of the British system'.

As an EU national Mario has advantages. 'Since 1996 it's been different, in that regulations changed in January 1996, and since then it's easier for, at least on paper, for movement freely in the EC [sic]. This means that titles obtained in other EC [sic] countries can be used when formally applying for jobs in the UK. It doesn't mean one gets the job. I managed to get a job [initial], which for me is an extension of surgical training which I felt I could not have done adequately in Italy'. He also spoke about other doctors who are EU citizens, 'other EC [sic] doctors [I know] who come here for a variety of reasons consider staying. It goes very much into personal stories. I know some who are not sure about staying ... they are planning [their careers]. And I know some who have gone back [to Italy]'. He also said that, 'less and less is the language [a problem for him]. I still find sometimes that language can become a small problem ... with people with

a thick regional accent ... a slight problem when they use very local variations of conversation'.

Mario and Ali's career success following migration contrasts with that of
the two final migrant doctors cited in this section. Both migrated from the
Punjab, India to Canada. Jaswinder,[17] his wife and son arrived in Canada in
1999 where he hoped, 'to practice medicine [and secure] a better future for
myself and my family'. He began to seriously consider migration using his
'portable' skills in the early 1990s, relying first on information from colleagues: 'my senior doctors were planning to move to Muscat [and Oman].
Two of them went there. I applied to the Ministry of Health [Oman], they
selected me as a doctor but I decided not to go for two years. [Then] my father
expired [died], after that I applied for information from other countries,
Australia, but I did not qualify. Then somebody told me about Canada, he
inspired me and I applied'. Once he put the Oman job 'on hold' he made use
of his wider social network of friends from medical school. One friend in the
UK discouraged him from trying in the UK; 'he told me it was hard'. 'I
decided to come to Canada because there would be better opportunities [than
in Oman]'. Migration flows from the Punjab to Canada are very strong
(Walton-Roberts 1998). Jaswinder reported 'about 50 people from my home
came [to Vancouver]'. So when he and his family arrived at Vancouver airport, 'my wife's aunt' gave them accommodation and helped him network. 'I
landed on the Sunday and on the Monday I began trying to work'.

Jaswinder has family members in a number of locations in Canada
(Winnipeg, Toronto, Prince George and Vancouver) as well as a cousin in
the USA. His wife has relatives in the USA, Australia and New Zealand.
He also has medical friends in the USA, 'six in different places (Michigan,
Phoenix, Detroit). One of my friends is in England, he is doing the PLAB[18]
and agency work. He wants to come to the USA. He is coming in June
[1999] to take the test'. But his work experience and qualifications do not
allow Jaswinder to practice medicine in Canada immediately (as indeed
elsewhere, see Figure 8.1, p.105) and opportunities are becoming tighter.
He estimates that of the 150 students with whom he trained 'about 40 to
50 are now abroad, mostly in the US. It was easier up to 1994 to get residency [in the USA], but the last three years [interview in 1999] there are
about 19,000 got turned down for residency [in the USA]'.

Despite having so many members of his family in Canada, at the time of
the interview Jaswinder was feeling that Canada had been a poor choice for
him: 'I think I have chosen the wrong country because wherever I go I am
more qualified that most doctors, but unqualified for the job in Canada'.
Since arriving he has decided not to try to get professional accreditation in
Canada but to aim for US accreditation, even though, 'the competition is
very tough'. He knows that once he has gained recognition, 'once you start
working there are other barriers. The racism is more visible'.

Pallab[19] also gained Indian medical qualifications in 1979 and then began
work. 'I was medical officer in charge of a rural hospital for two and a half

years before coming to England. I have relatives all over [the UK], Nottingham, London, Luton'. It was during his first visit to the UK that he met his future wife, the marriage arranged with the help of the UK relatives. Pallab returned to the UK in 1983 to marry her and they began to build a life together there. 'She was a pharmacist from Birmingham. My wife's family are all professionals'. He tried to get UK medical registration on a number of occasions, 'it was 1985 when I tried, I passed the theory and clinical but I did not score enough in the practical part'. After a failed second attempt he started applying for non-medical jobs, 'I applied for one, which was at an Indian Community Centre in ... [city in England]'. Then in 1991 the family moved from the UK to Canada. 'My wife's uncle [in Vancouver] was the motivation. He thought I should come and try [to practice medicine] here [in Vancouver]'. In Canada he, 'met many doctors and asked them what to do'. Following their advice he sat the US exams and failed. 'In 1994 I made up my mind that I'm not trying any more. So I started looking for [a non-medical] position. At the back of your mind you are spending money, taking a test is [Canadian] $1,400 to $1,600 each time'. Reflecting on life in Canada he said, 'the land of milk and honey doesn't happen. Materialistic things are what people have here [Canada] ... we want to give the best for our child'. As some of these case studies have illustrated, while medical qualifications can lead to secure employment in some countries, this is not always the case. Careers in medicine are highly regulated both by the state and by professional bodies, and securing professional registration is extremely difficult in some countries even those with shortages of doctors.

Caring for the world: overseas nurses

Nursing has long been regarded as, 'the most extreme example of the influence of gender on occupation choice; the classic case of "women's work" ' (Thornley 1996: 162). While the majority of nurses spend their working lives working in local care establishments in one local labour market (Ludingsen 1993: 22), there is a long history of international nurse mobility, with highly complex flows. Despite their importance (or because of it) many countries are experiencing a shortage of nurses (Figure 8.1; Buchan *et al.* 1994: 460; Sawada 1997), and these shortages are often filled by recruiting staff – mainly single women – trained in other countries. Today Filipina nurses work in many countries including Taiwan, where they accounted for 93 per cent of labour importation permits issued as of May 1995 (Tsay 1995: 186). They form an extremely high proportion of the nursing workers in private homes (99 per cent), as housemaids (93 per cent) and as nursing workers in institutions (95 per cent) (ibid. 185). They remit income back to the Philippines. Some migrant nurses leave partners and children behind while they fulfil their fixed term overseas contract (Hochschild 2000; Momsen 1989). Because of the sheer scale and complexity of nurse migration I will focus on the personal migration stories of non-UK qualified nurses to the UK and UK-qualified nurses. First, however, I will review issues connected with the immigration of nurses. While the focus of the book is dual career households this section largely examines 'solo' nurses, because permission to enter a country and nurse tends to be granted for the nurse and not her household.

As Pesman (1996: 2) notes in her account of the international migration of Australian women, 'in the 1950s and 1960s, there were only three acceptable reasons for leaving the nest: marriage, nursing or travelling overseas'. Before the age of mass travel, many single women used their 'portable' nursing skills to finance international travel and extended periods of stay in Europe, as is still the case today. The same is true for New Zealanders (Chant and McIlwaine 1998).

In the UK the state funded NHS has relied on overseas nurses at various points since its establishment after World War II. Nurses from the Caribbean migrated to the UK in the 1950s and 1960s, often in response to job advertisements placed in local newspapers. In 1960, for example, a severe shortage of nurses led the Ministry of Health to approach the Government of Barbados (Akiansanya 1988). Many migrant nurses tended to be 'channelled' into a non-career grade (enrolled nursing) but despite this, they tended to stay and build a career in the UK (Hardill 1998a).

In addition to West Indian nurses, Irish women formed a second group who were encouraged to migrate to train and/or work in the UK and elsewhere in the post-war period. Bronwen Walter (1989) undertook an extensive survey of Irish nurses working in London. She reported 1981 Census of Population statistics, which revealed that 17 per cent of Irish-born women living in London were in medical services (compared with eight per cent of white UK-born, and eight per cent of women in Eire) (ibid. 71). She interviewed nurses working in east and west London, and recorded that 90 per cent of the women came to the UK either to train, for career advancement or because of job availability (ibid. 75). They had responded to press recruitment campaigns in Ireland. Most felt that they had been 'pushed out' of Ireland by lack of nursing opportunities (ibid. 76).

Today overseas-qualified nurses are recruited to work in individual hospitals; about half of them gain posts through specialist recruitment agencies while the remainder are recruited directly by individual hospitals (with an international profile and contacts abroad). Some agencies are proactive in targeting UK hospitals with no international contacts of their own and staff shortages, offering them overseas-qualified nurses (conversation with an East Midlands hospital). Overseas-qualified and UK-qualified nurses are offered fixed term contracts of six months to two years duration. Traditionally overseas nurses (especially from the former Dominions) sought working holiday experience especially in London, whereas a number are now seeking posts offering professional development (Chamberlain Dunn Associates 1999: 14). Nurses migrate either as 'trailing spouses' or as solos; the latter are truly skilled transients as most move for a fixed term appointment, often for 'OE'. Some solos leave dependants behind, as is often the case with Filipana nurses (Tyner 2000).

Recruitment agencies headquartered in the UK and Ireland (maintaining websites) specialise in filling vacancies in the UK, Europe and the Middle East. These agencies advertise abroad for one specific hospital and deal with the selection of potential nurses for interview. They also organise United Kingdom Central Council (UKCC) registration or the verification of UK qualifications for the Middle East, and obtain work permits/visas if necessary as well as booking flights and accommodation en route to the hospital. They charge a fee per nurse. As a recent report (Chamberlain Dunn Associates 1999: 18) demonstrates, several small hospitals on the south coast of England considered using an agency but decided against it because of the costs and worries about issues of retention and work permits. It takes about three months for UKCC to process registration documents, so it is difficult for UK hospitals to use overseas nurses as an immediate solution to a nursing shortage.

Non UK-qualified nurses registered with UKCC provide an illustration of the comparative importance of non-UK countries as a source of 'new' nurses. As Table 8.3 clearly indicates, overseas-qualified nurses now account for 16 per cent of new admissions to the UKCC register. The proportion of nurses entering under EU regulations has increased during the 1990s (Table 8.4). Today one third (1,439) of the admissions are via EU arrangements (including 402 from Finland, 348 from Ireland, 171 from Sweden and 146 from

Table 8.3 Admissions to the UKCC register, 1984/5–1999/2000

Year	UK	EU/non-EU	Total
1984/5	34,056 (93.0%)	2,587 (7.0%)	36,643 (100%)
1990/1	29,755 (90.0%)	3,313 (10.0%)	33,068 (100%)
1993/4	29,885 (93.0%)	2,258 (7.0%)	32,143 (100%)
1997/8	22,132 (84.0%)	4,333 (16.0%)	26,465 (100%)
1998/99	21,901 (81.3%)	5,033 (18.6%)	26,934 (100%)
1999/2000	21,327 (74.2%)	7,404 (25.7%)	28,731 (100%)

Source: adapted from UKCC (2001)

Table 8.4 Admissions of non-UK qualified nurses to the UKCC register, 1990/1–1997/8

Method of admission	1990/1	1993/4	1997/8
EU arrangements	813 (24.4%)	456 (20.2%)	1,439 (33.2%)
Other	2,518 (75.6%)	1,802 (79.8%)	2,894 (66.8%)
Total non-UK qualified	3,331 (100%)	2,258 (100%)	4,333 (100%)

Source: adapted from UKCC (1998)

Germany) and two-thirds (2,894) are from elsewhere (including 1,170 from Australia, 472 from New Zealand, 393 from South Africa, and 203 from Canada (UKCC 1998).

As Table 8.5 reveals, the number of nurses registered with UKCC seeking verification to nurse abroad peaked during the economic downturn of 1990/1991. It should be borne in mind that these figures measure intention to nurse abroad. While the proportion seeking verification to nurse within the EU has increased over the period 1984–5 to 1997–8, the preferred destinations remains the former Dominions and the USA. In 1998, 16 per cent of verifications were requested for the EU while 69 per cent were requested for the former Dominions (1,329 for Australia, 143 for Canada and 549 for New Zealand) and 320 for the USA. Only 60 were issued for Saudi Arabia (UKCC 1998).

The requirement for language capabilities to communicate with patients and colleagues is likely to act as a barrier to international migration. Personal constraints on mobility (such as a partner and/or dependants) may also shape nurse's mobility (Hardill and MacDonald 1998). As Figure 8.3 illustrates, a number of socio-economic barriers may impede the international migration of nurses in pursuit of a medical career, but nurses do migrate, and for a number of reasons (Figure 8.4). In a recent study of British nurses working abroad, the most commonly quoted reason for going was to gain non-UK nursing experience (28 per cent); one quarter reported that their main reason for nursing abroad was that they had moved with their partner/family. A further 16 per cent indicated that it was a 'working holiday'[20] (especially to the USA and Australia) and 14 per cent revealed financial reasons (especially amongst those who had worked in the Middle East and Saudi Arabia) (Buchan *et al.* 1997).

Overseas nurses: migration and aspirations

Nurse recruitment and retention problems seem to be universal, and as I write this book, crises in the nursing profession appear almost every day in

Table 8.5 Number of documents issued to other regulatory authorities to verify registration with UKCC

Year	EU (including Ireland)	USA, Australia, New Zealand and Canada	Total
1984/5	362 (11.5%)	2,051 (65.1%)	3,150 (100%)
1990/1	1,217 (14.1%)	6,609 (76.6%)	8,626 (100%)
1993/4	1,100 (30.2%)	1,681 (46.2%)	3,639 (100%)
1997/8	544 (15.9%)	2,338 (68.8%)	3,400 (100%)
1998/99	719 (18.7%)	2,459 (64.2%)	3,830 (100%)
1999/2000	827 (16.2%)	2,872 (56.5%)	5,083 (100%)

Source: adapted from UKCC (1991; 2001)

- differences in the organisation and structures of health delivery systems
- variations in nurses' pay and grading
- variations in nurses' roles, responsibilities and position in the clinical hierarchy
- differences in career paths and promotion systems
- interruptions to careers and the lack of recognition given to overseas nursing experience
- access to continuing professional development
- perceived social and economic standing of nurses
- language barriers
- domestic commitments
- European Social Charter
- economic conditions/employment prospects

Figure 8.3 Barriers to international mobility of nurses
Source: IMS cited in Seccombe *et al.*, 1993, p.138

'permanent' move
the economic migrant: attracted by better standard of living
the career move: attracted by enhanced career opportunities
the migrant partner: unplanned move, as a result of spouse/partner moving

temporary move
the working holiday: nursing qualification used to 'finance' travel
the study tour: acquisition of new knowledge and techniques for use in home country
the student: acquisition of basic or post basic qualifications for use in home country
the aid worker: providing care or transferring care skills to a host country

Figure 8.4 The 'mobile nurse' typology
Source: IMS cited in Seccombe *et al.*, 1993, p.138

the press (see Figure 8.1, p.105). I begin this section with an analysis of overseas nurse recruitment with a case study of a hospital in a UK provincial town (Hardill and MacDonald 2000). Overseas recruitment is not targeted specifically at EU countries. The sole criteria for advertising for nurses in a country is that the recruitment agency appointed by the hospital thinks there is a surplus of trained nurses who they feel could be enticed to move as solos for a limited period to nurse in the UK. For this hospital, the 'preferred recruitment route' for those not holding UK nursing qualifications is through the recruitment agency, as it deals with the necessary paperwork demanded by UKCC to secure UK registration (key actor interview).

The current overseas recruitment programme for nurses commenced in 1997 when the hospital encountered difficulties in filling vacancies with newly qualified staff or with a national advertisement campaign. As Table

8.6 illustrates (see p.128), a small number (16) of overseas nurses are currently working in the case study hospital. This is much lower than would be found in hospitals in major metropolitan centres, especially London which have a much higher proportion of their nursing staff with overseas qualifications. The hospital's Director of Nursing liaises with the recruitment agency. The agency advertises for nurses in a specific country, which they choose, and prepares a shortlist for interview. The agency screens nurses for ability to speak English, qualifications and 'suitability' (essentially an ability to migrate as solos). A member of the hospital staff undertakes the interviews (in that country) or an interview may be by telephone. The necessary paperwork (UKCC registration, work permits/visas) is arranged by the agency. They also organise travel to the UK for the successful nurses. In theory, someone is at Heathrow airport to meet the migrant nurse, but this does not always happen. For these services the agency charges a fee of about £2,000 per nurse. The nurses are offered a one-year contract, and if s/he wishes to renew his/her job contract the agency charges a further fee.

In 1997, the agency began recruiting staff from South Africa (see Table 8.6, p.128). In early 1999, Canada was targeted, but as too few Canadian nurses responded to the advertisement, no interviews took place. The agency has since turned to Germany and nurses (qualified and student nurses) from all over Germany (and Poland) responded. The agency has set in motion the necessary registration procedures for seven nurses. At the time of writing only one nurse, from the former DDR, had taken up a post in the hospital, and another is expected. So despite press campaigns in South Africa, Canada and Germany during 1999 only six nurses are in post. In addition to these nurses there are Leonardo programme[21] nurses from Finland and two 'trailing spouse' nurses in international migrant households (from the USA and Australia).

The Finnish nurses come to the hospital as part of an EU-funded student mobility programme, the Leonardo Project. The Project is an exchange programme for qualified Finnish nurses who have been unable to find work in Finland where there is currently a surplus of trained nurses, a situation which will continue for a number of years. They are placed in a hospital in the UK or Germany for three months in groups of around eight at a time. They receive language and 'acculturation' training before departure and while in the UK the nurses receive 'pocket money' and accommodation. UKCC registration is organised through the Leonardo Programme and some delays have been experienced because of UKCC procedures so that some of the Finnish nurses have been unable to work as qualified nurses, and have had to work as care assistants.

All the interviewees in the current cohort have received their UKCC registration, enabling them to nurse, however the Leonardo organiser complained that the UK registration was 'more difficult' to acquire than the German. Indeed, in Germany, the Finnish nurses are able to work straightaway (key actor interview). The Assistant Director of Nursing in the UK

Table 8.6 Overseas nurses recruited to work at the case study hospital

Country of qualification	Key reason for considering nursing overseas	Length of contract	Method of recruitment	Number of nurses
Finland	As part of EU-funded training programme	3 months	Leonardo course run in Finland	8
Germany	Problems of health sector in Germany	One year, some want longer contracts	Agency	9 offered posts, of which 7 accepted, but only one one currently employed and one more is expected to come.
USA	Moved as 'trailing' spouse to UK	3 months trial period	Applied directly to case study hospital in response to UK job advertisement	1
Canada	Problems of health sector in Canada	One year, but has been renewed	Contacted agency herself while in Canada	1
Australia	Moved as 'trailing spouse' to UK	3 months trial period, now permanent night	Applied directly to case study hospital in response to UK job advertisement	1
South Africa	Overseas experience	One year	Agency	4

Source: Midlands UK hospital study (Hardill and MacDonald 2000)

agreed with this, saying: 'unfortunately I think that we have entered Europe [sic] but the UKCC is a little slow in recognising this. Instead of people having free passage from Europe straight into nursing here, they still have to put all their details in to the UKCC and get permission to register here, they don't have to do any further study but there's a laborious paperwork process and they are in the EC – which I find ridiculous!'

Most of the overseas nurse recruits are single women in their mid to late twenties. South African nurses, by contrast, tend to be older and often leave their families in South Africa. All those recruited are black. One German nurse who had a small child was not offered a post.[23] The nurses are motivated to come to the UK essentially because of job shortages or dissatisfaction with the hospital system at home. For example, one Finnish nurse (in her early twenties) said she came because: 'there are so few vacancies. It is very difficult [in Finland]'. A Canadian nurse in her late twenties gave similar reasons: 'when I started nursing in 1992, it was bad. We could find jobs in doctor's offices, but no hospital jobs. I had a choice between here [UK] and the States. There are many jobs in the States. To go to the States was not very exciting for me. [I] was looking through the paper one day and [UK jobs] were advertised [by] the agency that I came through. [I came to the UK to] travel and gain some work experience'. As this nurse explained, the Canadian labour market prompted her to consider jobs abroad, but the decision to come to the UK rather than go to the USA was for 'OE'.

For one South African nurse (early forties) 'OE' has also been an important element in her decision to come to the UK: 'The only place I could [travel and nurse in the apartheid era]... was in Namibia'. She returned to South Africa in 1992 prompted by her desire to vote for the first time in her life. But she left again in 1998: 'There are agencies in South Africa for nurses who want to go abroad. Most go to Saudi, then the UK. Saudi, I would only have gained financially. But in the UK [I thought] it would be quite challenging. I have always loved to come and visit First World hospitals. To compare things and get more knowledge and what is happening in nursing'.

Another South African nurse (early forties) also talked about 'OE': 'I know where I come from nobody has been to where Shakespeare was born'. Similarly, a German nurse (early twenties) commented: 'I always wanted to work in a foreign country before I was employed by the agency. I came here to get experience'. The overseas nurses (those recruited through agencies as well as the 'trailing spouses') tend to fill lower grade vacancies in the case study hospital. Some have experienced downward occupational mobility, a trend noted in other research on the impact of migration on nursing careers (Ackers 1998).

As with migrant nurses to the UK, most UK-qualified nurses migrate for relatively short periods. Lucy[23] for example, has spent most of her nursing career in the English Midlands apart from a short spell nursing in the USA, where she migrated as a solo: 'my parents moved [to Oregon, USA] in September 1979, and I started nursing in January 1979 ... I didn't want to go which really upset the apple cart. In the Northwest [USA] in the late [19]70s folks were still walking about in cowboy boots and denims. That's the reason I wouldn't move with my parents. It's all changed [there now], it's quite cosmopolitan now'. Lucy did migrate in the mid 1980s after a period nursing in two hospitals in the same town in the Midlands, ' I got a bit disillusioned here [Midlands, UK]. My mum was an enrolled nurse [non career grade] over here [UK], she's not sat any exams since she's been out there [USA]. She is working as a care assistant, and was earning far more than me. That's what prompted me to go'.

Thus for Lucy the move was the result of a combination of factors, disillusionment with nursing and the low pay in the UK. During the 1980s, a radical reorganisation occurred within the NHS embracing management styles, hospital cultures and the nursing career structure (for a fuller account see Halford *et al.* 1997; Walby *et al.* 1994.) For Lucy, a lot of the satisfaction and pride she had found earlier on in her nursing career had gone. Under these circumstances, given her first-hand knowledge of nursing opportunities in the USA, she decided to migrate to work there as a nurse.

Lucy's story is echoed in an account of working in the USA by Amanda Wilcock (1996: 16). She had worked in hospitals in Liverpool, Manchester and at St Mary's and Barts in London before moving to the USA. But her over-riding memory of these experiences was, 'of a service crippled by cutbacks. Nursing wasn't a smooth process ... financially as well as professionally my colleagues felt unappreciated for what they were doing and for our level of skills and knowledge' (ibid. 16). Low morale in her job matched by her low income pushed her to look abroad for work. In the USA she feels that her quality of life and career prospects are brighter, 'nurses are respected as professionals. They are paid accordingly and also receive the support to allow them to do their jobs' (ibid. 17).

The migration of UK-qualified nurses abroad has featured in several recent press reports. The death of Australian nurse Yvonne Gilford in Saudi Arabia and the subsequent trial of British-qualified nurses Deborah Parry and Lucille McLauchlin revealed that qualified nurses were the only foreign women allowed to work in Saudi Arabia (Harding 1997). Though employment there is financially rewarding migrant nurses to Saudi Arabia are often described in a devalued language, for example as 'screwing fodder' (ibid.).

In an account of nursing in Saudi Arabia written following the completion of a six month contract as a nurse there, Sally Robinson (1995: 47) commented that there were no Saudi nurses and that most ancillary, paramedic and support staff were foreign, although she did meet one female Saudi doctor in Riyadh. Australian, Filipina, Egyptian, Finnish and British nurses worked together, although pay scales for the nationalities differed, with the Western nurses being paid considerably more than Asian nurses were. The wages were calculated from baseline salaries in the nurses' home country, though they all earned more than they would have done back home. Not only did salaries vary by nationality but accommodation did also. Asian nurses were more likely to share rooms than Westerners (ibid. 47; see also Hardill 1998). Thus in Saudi Arabia overseas-qualified nurses have special salary scales dependent on their country of qualification, in contrast to overseas-qualified migrant nurses to the UK or USA, where they follow the salary scales of locally-qualified nurses.

Conclusion

In this chapter I have raised the visibility of skilled professional migration by revealing that specialist training acquired over several years by doctors and nurses facilitates their international mobility. This spatial mobility is highly regulated, however. As Kofman (2000) has noted, research on skilled migration has rather narrowly emphasised transnational corporate moves, which, especially in their higher ranks, remain male-dominated. I have offered insights into the factors taken into account by doctors and nurses relating to where to live and work, highlighting the ways in which labour markets, recruitment agency and visa and settlement regulations influence their migration flows. Adopting a biographical approach illuminates the complex interweaving of economic and non-economic factors involved in individual decision-making (Li *et al.* 1995). While economic factors were indeed instrumental (for example, the scarcity of job opportunities at home), and they were not the only considerations. The desire to travel and see the world, family contacts and educational benefits (work experience and language development) all featured as reasons influencing the decision to move and the choice of country. These non-economic reasons can outweigh personal downward occupational mobility and stifled personal aspirations that may result from international migration.

Other empirical studies on skilled labour migration have recognised the prevalence of such downward mobility in the context of returning professional expatriates, including medical workers (Hardill 1998; Hardill and MacDonald 1998). While some doctors can and do sustain a career in the host country, others experience barriers to upward social mobility (Anwar and Ali 1997). In some ways, nurses formed a 'reserve army' of labour, a temporary stopgap to fill labour shortages. The fact that nurses are 'encouraged' to migrate as solos and that their household situation is not deemed important by the receiving hospitals, is a further illustration of their 'marginality' (Hardill and MacDonald 2000).

9   Dual career couples and the new economy

In this book I have focused on a privileged group of households, dual career households, who are currently working in Canada, the USA and the UK, drawing on the personal and household biographies augmented by published statistics. While within migration studies there has been a plea for the use of a biographical approach (Halfacree and Boyle 1993), there is also a debate as to whether couples should be interviewed together or separately (Valentine 1999). We deliberately chose to interview couples separately in order to gain two perspectives on their lives together (Pahl 1989) as well as identifying sources of conflict and negotiation in gender role identities and divisions of labour beyond the 'milestone' events such as marriage and childbirth. In many instances the accounts of their lives together were told in very different ways.

I have used the household biographies to highlight the ways in which social mobility (career advancement) and spatial mobility (geographic migration) intertwine for them. The pursuit of a career begins early, through the acquisition of economic, cultural and social capital with the foundations laid through parental choice of school, in the choice of a specific university, and the choice of a specific course of study. The analysis of social and spatial mobility has been undertaken from a range of perspectives, and at a number of spatial scales, and has illustrated the ways in which place, race and gender intersect to make social and spatial mobility a very different experience within and between households. This is vividly illustrated by the overseas-qualified staff working in hospitals in Saudi Arabia. They are recruited globally but salary scales are dependent on their country of qualification, with Western staff receiving better terms and conditions of employment than staff recruited directly from Asia or Africa, and doctors are more highly valued than nurses. Migration is often conceived of as the outcome of a conscious rational decision but it can also be considered as a haphazard, non-rational, non-economic decision. Migration decisions are more complex and non-economic factors relating to the household as well as the individual careerist can also play an important role in migration decisions.

Combining career development and family life has arguably become ever more complicated in the context of perceptions of a breakdown of employment security in the 'new economy'and the rise of dual career households. Mobility (that is the social nature of movement) and migration are the 'markers of our time' (Said 1994). There is a blurring of business travel with commuting, residential mobility and migration and business travel and migration. This increased mobility and connectedness has produced 'landscapes of mobility' (Auge 1995: 44). Prior to destruction of the World Trade Centre, New York on 11 September 2001, four million people took international flights daily (Doyle and Nathan 2001: 1). The internationalisation of labour markets and the international dimension of business travel were sadly illustrated by the list of the nationalities of those who died in the destruction of the Twin Towers. The list of fatalities included passport holders from every continent, and included people who did not regularly work in the Twin Towers, but who were there in restaurants and offices attending business meetings.

The work–life/home balance becomes even more difficult to manage when one member of the household is faced with the prospect of relocating his/her workplace for career development reasons. This is the case whether the move is an internal one with their existing employer or a move to a new employer in a different location. Employers who demand from their managers and professionals 'commitment' as expressed in working above contracted hours, in travel at short notice, inter-regional mobility or skilled transience, neglect the human consequences – on individuals and their households – at their peril. It is in the interests of employers and employees that issues relating to work–life balance as well as the costs and benefits of relocation and other types of geographical mobility are fully understood, from both household and organisational perspectives, and the negative impacts minimised. Relocation for example is not just about changing people's jobs; it is about changing their lives.

Legislation addressing equal opportunities in the labour market has begun to tackle gender issues within the workplace, especially benefiting women managers and professionals, but as we have seen in this book, dual career households who prioritise both careers are not the norm. Moreover in many dual career households combining work and home is accommodated not by partners sharing the tasks of social reproduction but by commodification, with some tasks off-loaded onto non-household members, especially women (Hochschild 2000; Momsen 1999).

Recent trends in mobility suggest that less emphasis is being placed on relocation, and more on the alternatives of 'commuting assignments' and 'virtual working', involving working away from home for all/part of the week (CBI Employee Relocation Council 2001) – in both corporate strategies and individuals' career development strategies. These newer types of geographical mobility, sometimes on an international basis, pose new complexities and different challenges for dual career households. The reality of living in the new economy can mean living together 'apart'.

Notes

1 Social and spatial mobility in a global economy

1 But Goldthorpe did not include the self employed.
2 But in studies in the UK there appears to have been a break in the absolute rate of social mobility, especially upward social mobility, around 1979–81 (Aldridge 2001).

2 Households, careers and decision-making

1 Theories which are appropriate for traditional male breadwinner/female homemaker or heterosexual dual earner households with a conventional division of labour (where the male partner's career/job would be expected to take precedence over the woman's job, if she had one) need to embrace the increase in female labour force participation and their penetration into managerial and professional jobs, as well as changes in household arrangements, if they are to be relevant for dual career households.
2 Pete (manager with a large manufacturing organisation) and Helen (nurse) both in their mid forties with two grown up children, met when they were students. Pete's career has always led and they have made five inter-regional moves.
3 Suzanne (hospital consultant) and Ali (hospital consultant) are in their forties and met at medical school and have three children. They both have successful medical careers.
4 James (self-employed consultant) and Sarah (university researcher) are both in their early fifties, have two grown up children, and have lived together for over 30 years. They met at school, they went to the same university (he was a couple of years ahead of her). His job has always 'led' and as a result they have moved inter-regionally four times.
5 Pete (manager with a large manufacturing organisation) and Helen (nurse) both in their mid forties with two grown up children, met when they were students. Pete's career has always led and they have made five inter-regional moves.
6 Pam (paediatric ward sister) and Dave (nursing lecturer) are in their early forties and have been together since college days in London and have two children in their early twenties, who still live with them. Dave's career has always been prioritised, and determines where they live.
7 Pete (manager with a large manufacturing organisation) and Helen (nurse) both in their mid forties with two grown up children, met when they were students. Pete's career has always led and they have made five inter-regional moves.
8 Francine (associate professor) and Jean (development consultant) are in their forties and have one daughter aged six years. They are based in Quebec. They

have lived together for over a decade. His work tends to take him to Africa, and he has about four assignments there a year.

9 Beky (associate professor) is in her mid forties and Peter (professor) in his mid fifties have been together for 4 years. Both partners have been married before. He has grown up children and Beky a teenage son. He works on the west coast of North America and she in the east.

10 Hela (born Eastern Europe) and Teddie (born USA) are in their late forties, and for both it is their second marriage. Hela's undergraduate and Master's degrees are from Europe and she migrated to marry her first husband and completed her PhD in the USA.

11 Dan (air force pilot) and Kerry (former pilot, now a research worker) are in their early thirties and both trained as fighter pilots in the Canadian air force. Kerry only completed one tour and left, 'to do something else and rekindle some of my dreams'. They are a commuter couple; he works at an airbase two time zones from Kerry's place of work. They travel to see each other every four to six weeks.

12 Simon (a sales executive) and Samantha (a senior analyst programmer) are in their early thirties, and work full time for the same manufacturing company. They have been together for just over a decade, and met at work when they were both, 'on a fast-track general management scheme'.

13 James (self-employed consultant) and Sarah (university researcher) are both in their early fifties, have two grown up children, and have lived together for over 30 years. They met at school, they went to the same university (he was a couple of years ahead of her) and his job has always 'led' and as a result they have moved inter-regionally four times.

14 Diana (mid thirties, university lecturer) and Bob (mid forties, further education lecturer) met when Diana was a student. They have a baby son and while they prioritise her career, they are very much equal partners. They employ a nanny to care for their son. Diana has held a series of fixed term contracts but now has a tenured lectureship.

15 Becky (software consultant) and Gary (accountant) are in their early thirties and met at university. They have a baby son and while they prioritise his career, until her maternity leave they both worked very long hours.

16 Claire (senior social worker) and Ian (self-employed surveyor) are in their early forties and have been together since university days and have two teenage children. Ian's career has always 'led'.

17 Nigel (a part-time lecturer with contracts with three employers) and Anne (a school teacher) are both in their early forties, and have been together for ten years. This is Anne's second marriage. Anne's two children from her first marriage live with them along with their eight-year-old daughter. Anne has the 'lead' career.

18 Simon (a sales executive) and Samantha (a senior analyst programmer) are in their early thirties, and work full-time for the same manufacturing company. They have been together for just over a decade, and met at work when they were both, 'on a fast-track general management scheme'

19 The Blair household is also assisted by paid domestic help.

20 Jane (food technologist) and Mark (sales representative) are in their mid thirties and have been together for about a decade. They have a young child who attends a private nursery.

21 Joanne (doctor) and Ken (regional sales manager) are in their early forties and met about a decade ago when they both had established careers. Joanne has been married before. They have a five-year-old son at a private school and they employ a nanny.

22 The official definition of part-time work in the UK is 29 hours per week.

3 Defining status and success through the pursuit of a career

1 While it is true that many careerists feel that the world of paid work is uncertain, and that the idea of a job for life is not a reality, those individuals and households without paid work have the most uncertain future.

2 Bourgeois Bohemians.

3 Simon (a sales executive) and Samantha (a senior analyst programmer) are in their early thirties, and work full time for the same manufacturing company. They have been together for just over a decade, and met at work when they were both, 'on a fast-track general management scheme'

4 Pete (manager with a large manufacturing organisation) and Helen (nurse) both in their mid forties with two grown up children, met when they were students. Pete's career has always led and they have made five inter-regional moves.

5 Claire (40, born West Yorkshire, senior social worker) and Ian (42, born East Anglia, self-employed surveyor) have been together since university days and have two teenage children. Ian's career has always 'led'.

6 Beky (university lecturer) is in her mid forties and Peter (university professor) in his mid fifties have been together for four years. Both partners have been married before. He has grown up children and Beky a teenage son. He works on the West Coast of North America and she in the east.

7 Francine (university lecturer) and Jean (development consultant) are in their forties and have one daughter aged six years. They are based in Quebec. They have lived together for over a decade. His work tends to take him to Africa, and he has about four assignments there a year.

8 In Quebec the university sector is divided between French and English speaking institutes.

9 James (late fifties) is a self employed consultant married to Sarah (mid fifties), who is a part-time university researcher. They have been together for over 35 years and have two grown up children. His career has taken them from London to Lancashire to Essex to the East Midlands. He has been made redundant himself at least twice.

10 Jason (a research fellow) and Caroline (a scientific officer) are in their early thirties and they live in a fairly 'old' terrace in a small town centre in Warwickshire, central England.

11 Gerry (self-employed editor) and Jenny (self-employed information broker) are in their forties, have been together since university and have two daughters (aged eight and twelve years). They have lived in a small village in the English Midlands, and prior to that in London.

12 Before living in the Midlands he had worked in London, and they lived first in NW10, and then in Buckinghamshire, from where he commuted daily to the city (of London).

13 This may be changing in the case of the UK, Home Secretary David Blunkett is planning a radical restructuring of UK immigration policy.

4 Defining status and success through residential property

1 The Quiet Revolution (or *la Revolution Tranquille*) refers to a politically-instigated programme of social and cultural reform in the province of Quebec and was led by the Quebec Liberal Government, which came into power in the early 1960s. The Liberal Premier of Quebec during the Quiet Revolution was Jean Lesage, who was elected using the slogan, '*maitres chez nous*' (masters in our house).

2 It has been phased out in the UK by the administration of Tony Blair.

3 In Canada they are called grow homes.

4 Betty and Bill are in their mid forties and have been together for about 15 years. They have a seven-year-old son, Tom, who was born when they were in Saudi Arabia. Bill's career has always 'led' and Betty is the prime carer who has spent periods as a full-time homemaker. After university Bill trained as a quantity surveyor. He has worked in Great Britain and in Saudi Arabia. For the first two years of his assignment in Saudi they were a dual location household. They returned to Great Britain in 1992 and Bill, 'spent two years working in Great Britain [based in London] with the company I was abroad with' but then became self employed in order to try and control the amount of business travel he does. They live in a village in the Midlands.

5 The blurring of boundaries between home and work

1 It should be remembered feminists have challenged these strict divisions.

2 Simon (a sales executive) and Samantha (a senior analyst programmer) are in their early thirties, and work full time for the same manufacturing company. They have been together for just over a decade, and met at work when they were both 'on a fast-track general management scheme'

3 Joanne (doctor) and Ken (regional sales manager) are in their early forties and met about a decade ago when they both had established careers. Joanne has been married before. They have a five-year-old son at a private school and they employ a nanny.

4 James (self-employed consultant) and Sarah (university researcher) both in their early fifties, have two grown up children. They have lived together for over 30 years. They met at school, went to the same university (he was a couple of years ahead of her) and his job has always 'led'. As a result they have moved inter-regionally four times.

5 Jason (a research fellow) and Caroline (a scientific officer) are in their thirties and have been together for 11 years. They met at university when she was working as a research assistant and he was doing postgraduate work. They have both held fixed term contracts, and Jason has been unemployed twice.

6 Simon (a sales executive) and Samantha (a senior analyst programmer) are in their early thirties, and work full time for the same manufacturing company. They have been together for just over a decade, and met at work when they were both 'on a fast-track general management scheme'

7 Suzanne (hospital consultant) and Ali (hospital consultant) are in their forties and met at medical school and have three children. They both have successful medical careers.

8 A total of 11 per cent of the households who responded to the self-completion questionnaires contained one partner who had worked abroad either before or during their relationship (for a fuller discussion see Hardill 1998; Hardill and MacDonald 1998).

9 Joanne (doctor) and Ken (regional sales manager) are in their early forties and met about a decade ago when they both had established careers. Joanne has been married before. They have a five-year-old son at a private school and they employ a nanny.

10 Hela (born Eastern Europe) and Teddie (born USA) are in their late forties, and for both it is their second marriage. Hela's undergraduate and Master's degrees are from Europe. She migrated to marry her first husband and completed her PhD in the US.

11 Carrie (personnel manager) and Jack (customer services manager) are in their forties. Both had established careers when they met; they have been together for a decade. Her career leads; she earns more than he does. They have two young children.

12 Margaret (personnel manager) and John (senior manager) both in their late thirties, have two school age children. They met a decade ago when they had established careers. His job is prioritised.
13 The psychological barrier is longer in North America.
14 Pam (paediatric ward sister) and Dave (nursing lecturer) are in their early forties and have been together since college days in London. They have two children in their early twenties who still live with them. Dave's career has always been prioritised, and determines where they live.
15 Joanne (doctor) and Ken (regional sales manager) are in their early forties and met about a decade ago when they both had established careers. Joanne has been married before. They have a five-year-old son at a private school and they employ a nanny.
16 Jane (food technologist) and Mark (sales representative) are in their mid thirties and have been together for about a decade. They have a young child who attends a private nursery.
17 Joanne (doctor) and Ken (regional sales manager) are in their early forties and met about a decade ago when they both had established careers. Joanne has been married before. They have a five-year-old son at a private school and they employ a nanny.
18 Dual career households are more fortunate than other households in that they tend to possess the economic resources to commodify childcare (see Chapter 2).
19 James (self-employed consultant) and Sarah (university researcher) both in their early fifties, have two grown up children. They have lived together for over 30 years. They met at school, went to the same university (he was a couple of years ahead of her) and his job has always 'led'. As a result they have moved interregionally four times.
20 Frank (freelance editor) and Alison (information broker) are in their early forties and have two school age daughters.

6 Spatial mobility within the education system

1 Nearly one third of US H1B work visas are given to Indians, mostly IT specialists (Rao 2000).
2 New Zealanders wanting to pursue academic careers are strongly advised to gain PhDs from outside New Zealand.
3 This massive expansion has not necessarily been accompanied by an increase in the size of the 'graduate job market'.
4 President Kennedy established the US Peace Corps in 1961 to promote world peace and friendship. Currently 7,000 Peace Corps volunteers are serving in 77 countries. Their assignments last for two years in which they work with the local population on a range of projects embracing education, the local environment and creating economic opportunities (www.peacecorps.gov/about/index.html).
5 When universities did not award higher degrees.
6 Two other long established destinations for postgraduates are France and Germany (Tysome 1999).
7 Hela (born Eastern Europe) and Teddie (born US) are in their late forties. For both it is their second marriage. Hela's undergraduate and Master's degrees are from Europe. She migrated to marry her first husband and completed her PhD in the US.
8 Tania (33) is an academic married to Olav, a fellow Eastern European academic. They have a six-year-old son.
9 State schools in Canada and private schools in the UK are chosen because they teach the International Baccalaureate. These schools often advertise in Hong Kong (Pe-Pua *et al.* 1996)

10 Janice and Nicholas are in their late forties and have three children. She is a university lecturer and he a government official in West Africa. Their three children attend private schools in the UK.

11 Solomon and Sandy are in their early fifties and have five children. After gaining a PhD in the US he has had a management career with a multinational company. Sandy is a trained teacher. Their five children have been born in the US, the UK or Austria. Solomon is now working in West Africa while Sandy and the children live in the UK.

12 Ranji and her husband Monder are 40 years old and migrated from India to Canada in 1998. Monder came six months ahead of Ranji and their two teenage children. They were a dual career household, both employed as government scientists in India. They left secure, challenging jobs largely for their children to have access to Canadian universities as home students. When I interviewed Ranji and Monder, they were both following masters' courses, Ranji in GIS and Monder in forestry management to try and improve their job prospects in Canada, but Ranji had low expectations, 'everything was for the children'.

13 Joanne and Bob are in their early thirties, and have landed immigrant status in Canada. They migrated from the UK in the 1990s to Canada, before they decided to co-habit. Both had a desire to migrate and Bob had relatives in Canada. They have since married. Bob is self-employed running a small construction business.

14 Such households are few and far between in Japan. Hirohito and Norika (born in Japan) are in their early fifties and have three children. Like many professional women, Norika stopped her educational career when her children were born. Hirohito is a university lecturer. Their two daughters are both at university, one in the USA and the other in the UK (for a semester).

15 Barry (born UK) and Jenny (born USA) are in their forties. They met when they were both graduate students at a US university. Barry's career has led and determined where they live. He has a successful academic career and Jenny works in the arts. She has taken career breaks and has picked up her career again. They have two teenage children.

16 A number are amongst refugee and asylum seekers in the UK at the time of writing though their skills and expertise are certainly not 'valued' by the receiving country.

17 It was also noted that competitive pressures within the British academy, such as the research assessment exercise, might also encourage universities to employ staff from abroad (Swain 1999).

18 Within the academy sabbaticals originated in the USA in the late nineteenth century. They designate a period of leave from duty granted to university teachers at certain intervals (originally every seven years) for the purposes of study and travel.

19 Irene and Olav (born Eastern Europe) are in their late thirties. Both are university lecturers and are pursuing careers in their country of birth. They have no children and each has worked abroad for short periods of time, in both Europe and the USA.

7 Organisational careers: patriarchy and expatriate work

1 James (university lecturer) and Marie (pharmacist) (UK) are both in their forties and have two children, one still at school and one at university. They both work full-time and attach priority to his career. They have been together for over a decade since James left the military.

2 Dan (air force pilot) and Kerry (former pilot, now a research worker)(Canada) are in their early thirties. Both trained as fighter pilots in the Canadian air force.

Kerry only completed one tour and left, 'to do something else and rekindle some of my dreams. They are a commuter couple; he works at an airbase two time zones from Kerry's place of work. They travel to see each other every four to six weeks.

3 Hans and Bettina (USA) are in their late thirties and are German citizens who currently work in the USA for a German multinational. They have two children, a son aged 12 and a daughter aged 16 years. They have lived for four years in the USA and are about to return to Germany. They are not sure where they will be moving to as Hans is currently negotiating a suitable job with his German employer.

4 Jane (a food technologist) and Mark (a sales representative) (UK) are in their mid thirties and have been together for about a decade. They have a young child who attends a private nursery. Jane is currently working part-time because she was finding it too difficult to combine full-time demanding work as a food technologist with the added domestic responsibilities of a young child. Where they have lived has tended to be shaped by Jane's job as 'he was always on the road'.

5 Bill and Betty (UK) are in their mid forties and have been together for about 15 years. They have a seven-year-old son, Tom, who was born while they were in Saudi Arabia. Bill's engineering career has always 'led'. Betty is prime carer and has spent periods as a full-time homemaker.

6 Jenny (40) and Peter (36) (UK) have been together for a decade and have three young sons. Peter had a military career, and they met when he was retraining. His engineering skills secured him a post with an oil company. Jenny is a careers guidance officer, but took a career break when living in Saudi Arabia.

7 The life of lower skilled migrant workers from South East Asia in Saudi Arabia – another strata in internationalised labour markets – is totally different.

8 Susan and Rahul (UK) are both in their thirties. Although they both have medical careers, they give priority to his career. He is a consultant paediatrician and she a nurse tutor. They have been together since 1983 and have three school age sons.

8 Professional careers and skilled international migration: case studies of health care professionals

1 But for a minority of nurses migration flows have been long distance for a considerable time.

2 The authors felt that there was a correlation between the lower status afforded to nurses in Australia and Singapore and the fact that they were former British colonies/Dominions (Johnson and Bowman 1997: 205). Their nursing structures are modelled on those in the UK where nursing is regarded as a semi-profession.

3 This is also true in academia.

4 In Osler's opinion, 'there is no higher mission in this life than nursing God's poor' (Verney 1957: 67).

5 This is the terminology of the doctors themselves; their professional organisation in UK is called the Overseas Doctors Association. It was also used by Anwar and Ali (1987) in their report on overseas doctors for the Commission for Racial Equality in UK.

6 There has also been a considerable amount of illegal immigration of nurses from the Philippines to the USA (see Momsen 1999).

7 http://www.infophil.com/india/alumni

8 In 2000 about eight million people in the USA have healthcare provided by HMOs. An HMO is a group that provides medical care (outpatient and hospital services) to a group of individuals for a fixed periodic payment. They are for-profit corporations with stockholders.

9 The Labour administration of Tony Blair has recognised the scale of the shortage and announced that an extra 1,000 medical students per annum will be trained from the academic year 2000–1; at present 4,000 per annum are trained (Goddard 1999).

10 In a recent analysis of the class of 1998 from the Medical School at Bristol University, of the 140 who qualified, two have returned to their country of birth (Malaysia and Singapore); 25 have taken up posts as junior doctors in Australia, three in New Zealand and one in Hong Kong (Woodward 1999).

11 The General Medical Council is the regulatory body for the medical profession.

12 The EEA is the single market migratory space of the European Union plus the countries of the European Free Trade Area.

13 Sameera (39) and Mushtaq (41) were both born and educated in Pakistan and have been married for three years. They have no dependents. Sameera obtained an MBBS (Batchelor of Medicine, Batchelor of Science) at a medical school in Pakistan, came to the UK post-qualification for further education and then returned to work in Pakistan. She emigrated to the UK to marry Mushtaq.

14 Suzanne (45) and Ali (47) have been together since 1988. They have three children (aged eight, six and two years). Suzanne is Scottish and met Ali (Iranian) in 1982 when he came to undertake post-qualification studies in Scotland. Ali's parents live with them for periods of up to six months at a time 'they can only get six month visas'. Thus three generations live in the same household though Suzanne is prime carer. They employ an au pair to help with the childcare. Ali's father does not enjoy good health. Both Suzanne and Ali are hospital consultants (Suzanne in the East Midlands and Ali in the West Midlands) officially working 35 hours a week but effectively putting in, '45 to 50 with overspill'.

15 Mario (38), a hospital doctor born in Italy and Joanne (40) a full-time accountant born in the UK, met while Mario was working in the UK. They have been married for a couple of years.

16 Judy is the 28-year-old daughter of Sam and Victoria who are in their fifties. Sam is an engineer and Victoria a librarian. They met when they were undertaking postgraduate studies in the US in the 1960s. Judy was born in the USA when her Nigerian parents were postgraduates. She holds dual nationality, Nigerian and US. She qualified in Nigeria and like many of her class has migrated to the USA because of career blockages due to the unstable economy and gender issues.

17 Jaswinder and his dentist wife, Rajinder, are in their thirties with one son. Jaswinder had worked over for over a decade in a government hospital. They decided to migrate, 'for a better future'.

18 Language proficiency in English examination.

19 Pallab (trained doctor) and his wife Ranji (pharmacist) are in their forties and have a teenage son. Pallab was born in India and Ranji in the UK. Since marrying in the 1980s they have lived in the UK and Canada.

20 This was the term used in the report but perhaps it is more 'OE'.

21 An EU-funded mobility programme.

22 Overseas nurses are expected to come as solos. The only hospital accommodation available is in the nurses' home, and this accommodation is for solos. In contrast, doctors have access to hospital accommodation for both solos and 'families'. Migrant nurses can take their spouse to some Middle East countries providing s/he has also secured employment and a work permit.

23 Lucy is in her late thirties, and married for the second time. Her current husband is Aart, a Dutch-qualified doctor. They have two young children. She ' did two years orthopaedic training, after that I did my three years Registered General Nursing (RGN) qualification'. Aart works as a general practitioner in UK.

References

Abelson, R. (1998) 'Part-time work that adds up to a full-time job', *New York Times*, 2 November: A1–A18

Ackers, L. (1998) *Shifting Spaces: Women, Citizenship and Migration within the European Union*, Bristol: Policy Press

Adam, B. and Van Loon, J. (2000) 'Introduction: Repositioning risk; the challenge for social theory', in B. Adam, U. Beck, and J. Van Loon, (eds) *The Risk Society and Beyond: Critical Issues for Social Theory*, London: Sage 1–31

Adams, J. (1999) *The Social Implications of Hypermoblity*, Paris: OECD

Adler, N. (1994) 'Competitive frontiers: Women managing across borders', *Journal of Management Development*, 13: 24–41.

Adler, N. (1997) *International Dimensions of Organizational Behavior*, Cincinnati: South Western College Publishing Company 3rd Edition

Adonis, A. and Pollard, S. (1998) *A Class Act: the Myth of Britain's Classless Society*, London: Penguin

Akiansanya, J. A. (1988) 'Ethnic minority nurses, midwives and health visitors: What role for them in the National Health Service?', *New Community* 14, 3: 444–50

Aldridge, S. (2001) *Social Mobility: a Discussion Paper*, London: Performance and Innovation Unit

Allen, I. (1988) *Doctors and their Careers*, London: Policy Studies

Amin, A. (1994) *Post-Fordism: A Reader*, Oxford: Blackwell

Amrop International (1995) *The New International Executive: Business Leadership in the 21st Century*, New York: Amrop/Harvard University

Anderson, M., Bechhofer, F. and Kendrick, S. (1994) 'Individual and household strategies', in M. Anderson, F. Bechhofer and J. Gershuny (eds) *The Social and Political Economy of the Household*, Oxford: SCELI Oxford University Press 19–67

Antonopoulou, S. N. (2000) 'The process of globalisation and class transformation in the west', *Democracy and Nature*, 6, 1: 37–54

Anwar, M. and Ali, A. (1987) *Overseas Doctors: Experiences and Expectations*, London: Commission for Racial Equality

Arber, S., Gilbert, S. and Dale, A. (1987) 'Paid employment and women's health. A benefit or source of role strain', *Sociology of Health and Illness*, 7: 375–400

Arthur, M. B., Inkson, K. and Pringle, J. K. (1999) *The New Careers: Individual Action and Economic Change*, London: Sage

Assar, N. N, (1999) 'Immigration policy, cultural norms, and gender relations among Indian-American motel owners', in G. A. Kelson and D. L. DeLaet (eds) *Gender and Immigration*, Basingstoke: Macmillan 82–102

Auchincloss, L . (1987) *Diary of a Yuppie*, London: Weidenfield and Nicholson

Auge, M. (1995) *Non-places: Introduction to an Anthropology of Supermobility*, London: Verso

Bailey, P. and Culbert, L. (2001) 'Nursing grads tempted to head south', *Vancouver Sun*, 26 June: A4

Bailyn, L. (1993) 'Individual constraints occupational demands and private life', in L. Bailyn *Breaking the Mould*, New York: Free Press, Chapter 3

Baines, S. (1999) 'Servicing the media: Freelancing, tele-working and enterprising careers', *New Technology, Work and Employment* 14, 1: 18–31

Ballard, J. G. (2000) *Super Cannes*, London: Harper Collins

Bannister, C. and Gallent, N. (1998) 'Trends in commuting in England and Wales – becoming less sustainable?', *Area* 30, 4: 331–42.

Beck, U. (1992) *Risk Society: Towards a New Modernity*, London: Sage

Beck, U. and Beck-Gernsheim, E. (1995) *The Normal Chaos of Love*, Cambridge: Polity Press

Berk, S. F. (1985) *The Gender Factory*, New York: Plenum Press

Bjoras, G. (1989) 'Economic theory and international migration', *International Migration Review* 23: 457–85.

Blossfeld, H-P. and Drobnic, S. (eds) (2001) *Careers of couples in contemporary societies*, Oxford: Oxford University Press

Bogue, D. J. (1977) 'A migrant's-eye view of the costs and benefits of migration to a metropolis', in A. A. Brown and E. Neuberger (eds) *International Migration: A Comparative Perspective*, New York and San Francisco: Academic Press

Bondi, L. (1999) 'Stages on journeys: Some remarks about human geography and psychotherapeutic practice', *Professional Geographer* 51, 1: 11–24

Bonney, N. and Love, J. (1991) 'Gender and migration: Geographical mobility and the wife's sacrifice', *Sociological Review* 39: 335–48.

Booth, C., Darke, J. and Yeandle, S. (eds) (1996) *Changing Places: Women's Lives in the City*, London: Paul Chapman

Bourdieu, P. (1986) *Distinctions: A Social Critique of the Judgement of Taste*, Cambridge, MA: Harvard University Press

Bowlby, S., Gregory, S. and McKie, L. (1997) '"Doing Home"': Patriarchy, caring and space', *Women's Studies International Forum* 20, 3: 343–50

Bown, L. (2000) 'The give-and-take case for overseas students', *Times Higher Educational Supplement Opinion*, 23 June

Boyd, M. (1989) 'Family and personal networks in international migration: Recent developments and agendas', *International Migration Review* 23: 457–85

Brah, A. (1996) *Cartographies of Diaspora*, London: Routledge

Brannen, J. and Moss, P. (1988) *New Mothers at Work: Employment and Childcare*, London: Unwin

Breen, R. and Rottman, D. (1998) 'Is the national state the appropriate geographical unit for class analysis?', *Sociology* 32, 1;1–21

Breheny, M. (ed.) (1999) *The People: Where will they Work?* London: Town and Country Planning Association

Brewster, C. and Scullion, H. (1997) 'A review and agenda for expatriate HRM', *Human Resource Management Journal* 7, 3: 32–41

Brooks, D. (2000) *Bobos in Paradise: the New Upper Class and How They Got There*, New York: Simon and Schuster

Bruegel, I. (1996) 'The trailing wife: A declining breed? Careers, geographical mobility and household conflict in Britain 1970–89', in R. Crompton, D. Gallie and K. Purcell (eds) *Changing Forms of Employment: Organisations, Skills and Gender*, London: Routledge

Bryant, C. R. and Coppock, P. M. (1991) ' The city's countryside', in T. Bunting and P. Filion (eds) *Canadian cities in transition*, Oxford: Oxford University Press, 209–38

BT (1999) *Working from home*, Online. Available HTTP: http://www.workingfromhome.co.uk/wfh/homepage.htm (accessed 11 October 2000)

Buchan, J., Seccombe, I. and Ball, J. (1994) 'The international mobility of nurses: A UK perspective', *International Journal of Nursing Studies*, 31, 2: 143–54

Buchan, J., Seccombe, I. and Thomas, S. (1997) 'Overseas mobility of UK based nurses', *International Journal of Nursing Studies*, 34, 1: 54–62

Bunting, M. (2000) 'Taking the struggle out of family life', *The Guardian*, 28 September: 12

Butler, T. (1995) 'The debate over the middle classes', in T. Butler and M. Savage (eds) *Social Change and the Middle Classes*, London: UCL Press 26–40

Butler, T. (2001) 'Middle-class households and the remaking of urban neighbourhoods – some recent evidence from London', Paper presented at the Annual Conference of The Association of American Geographers, New York, March 2001

Cabinet Office (1998) *Women's Incomes over the Lifetime*, Online. Available www.womens-unit.gov.uk (accessed 11 October 2000)

Cabinet Office (1999) *Rural Economies: A Performance and Innovation Unit Report*, London: Cabinet Office.

Caligiuri, P. M. and Tung, R. L. (1999) 'Comparing the success of male and female expatriates from US-based multinational company', *International Journal of Human Resource Management* 10, 5: 763–82

Cannadine, D. (1998) *The Rise and Fall of Class in Britain*, New York: Columbia University Press

Carlisle, D. (1996) 'A nurse in any language', *Nursing Times*, 92, 39: 26–27

Carnoy, M. (2000) *Sustaining the Economy: Work, Family, and Community in the Information Age*, New York: Russell Sage Foundation

Carpenter, M. (1993) 'The subordination of nurses in health care', in E. Riska and K. Wegar (eds) *Gender, Work and Medicine: Women and the Medical Division of Labour*, London: Sage 95–130

Cassidy, J. (1995) 'Flying High in America', *Nursing Times* 92, 6: 16–17

CBI Employee Relocation Council (2001) 'Hot Topics', http://www.cbimobility.org (accessed April 10 2001)

Chamberlain Dunn Associates (1999) *Employing Nurses and Midwives: How to Recruit Nurses From Abroad*, Richmond, Surrey: Chamberlain Dunn Associates

Champion, T., Fotheringham, S., Rees, P., Boyle, P. and Stilwell, J. (1998) *The Determinants of Migration Flows in England: A Review of Existing Data and Evidence*, Newcastle upon Tyne: University of Newcastle, University of Leeds

Chant, S. and McIlwaine, C. (1998) *Three Generations, Two Genders, One World: Women and Men in a Changing Century*, London: Zed books

Choko, M. and Harris, R. (1990) 'The local culture of property: A comparative history of housing tenure in Montreal and Toronto', *Annals of the Association of American Geographers* 80, 1: 73–95

Christie, I. and Hepworth, M. (2001) 'Towards the sustainable e-region', in J. Wilsdon (ed.) *Digital Futures: Living in a Dot-com World*, London: Earthscan. 140–62.

Cloke, P. and Thrift, N. (1990) 'Class and change in rural Britain', in T. Marsden, P. Lowe and S. Whatmore (eds) *Rural Restructuring: Global Processes and their Responses*, London: Fulton

Cloke, P., Phillips, M. and Thrift, N. (1995) 'The new middle classes and the social contructs or rural living', in T. Butler and M. Savage (eds) *Social Change and the Middle Classes*, London: UCL Press 220–40

Cohen, R. (1997) *Global Diapsoras: an Introduction*, London: UCL Press

Commonwealth Secretariat (2001) *Migration of Health workers from Commonwealth countries: Experiences and Recommendations for Action*, London: Commonwealth Secretariat

Cooper, C., Liukkonen, P. and Cartwright, S. (1996) *Stress Prevention in the Workplace: Assessing the Costs and Benefits to Organisations*, Dublin: European Fund for the Improvement of Living and Working Conditions

Corden, A. and Eardley, T. (1999) 'Sexing the enterprise: Gender, work and resource allocation in self employed households', in L. McKie, S. Bowlby and S. Gregory (eds) *Gender, Power and the Household*, Basingstoke: Macmillan 207–25

Countryside Agency (2000) *State of the Countryside 2000*, Wetherby: Countryside Agency

Coupland, D. (2000) *Generation X: Takes from an Accelerated Culture*, London: Abacus

Coward, R. (1997) 'Why nurses are nursing a grievance', *The Guardian* 3 Feb: 15

Cox, T. (1978) *Stress*, London: Macmillan Education

Cresswell, T. (2001) 'The production of mobilities', *New Formations* 43: 11–25

Crompton, R. (1997) *Women and Work in Modern Britain*, Oxford: Oxford University Press

Crompton, R. and Harris, F. (1998) 'Gender relations and employment: The impact of occupation', *Work, Employment and Society*, 12, 2: 297–315

Crompton, R., Gallie, D. and Purcell, K. (1996) *Changing Forms of Employment: Organisations, Skills and Gender*, London: Routledge

Dahms, F. (1998) 'Settlement evolution in the arena society in the urban field', *Journal of Rural Studies*, 14, 3: 2

Darby, R. (1995) 'Developing the Euro-manager: Managing in a multi-cultural environment', *European Business Review* 95: 13–15

Davidoff, L. and Hall, C. (1987) *Family Fortunes: Men and Women of the English Middle Class 1780–1850*, London: Hutchinson

Davies, C. (1996) 'The sociology of professions and the profession of gender', *Sociology*, 30, 3: 661–78

Davies, J. and Deighan, Y. (1986) 'The managerial menopause', *Personnel Management*, March, 28–32

Daycare Trust (2000) 'Shift parents left without childcare', in *No More Nine to Five: Childcare in a Changing World*, London: Daycare Trust. Online. Available HTTP   http://www.daycaretrust.org.uk/NewsNewsDetail.cfm?NewsID=37 (accessed 11 October 2000)

Dean, H. (1998) 'Popular paradigms and welfare regimes', *Critical Social Policy*, 18, 2: 131–56

Delaet, D. (1999) 'Introduction: The invisibility of women in scholarship on international migration', in G. A. Kelson and D. L. DeLaet (eds) *Gender and Immigration*, Basingstoke: Macmillan 1–20

Delphy, C. (1981) 'Women in stratification studies', in H. Roberts (ed.) *Doing Feminist Research*, London: Routledge 114–28

DfES (2000) *Work Life Balance. Changing Patterns in a Changing World*, London: HMSO

Dixon, J. (1999) 'Canada or the United States: Which has the best social security system – and does it really matter?', *Canadian Review of Social Policy* 44: 71–83

Dormsch, M. E. and Ladwig, D. H. (1998) 'Dual earner families', *European Network: Family and Work*, 4/98 Erkath: European Commission

Doyle, J. (2000) *New Community of New Slavery: the Emotional Division of Labour*, London: The Industrial Society.

Doyle, J. and Nathan, M. (2001) *Wherever Next: Work in a Mobile World*, London: The Industrial Society

Doyle, J. and Reeves, R. (2001) *Time Out: the Case for Time Sovereignty*, London: The Industrial Society

Drew, E. (1998) 'Changing family forms and the allocation of caring', in E. Drew, R. Emerek and E. Mahon (eds) *Women, Work and the Family in Europe*, London: Routledge 27–35

Dumelow, C., Littlejohns, P. and Griffiths, S. (2000) 'Relation between a career and family life for English hospital consultants: Qualitative, semi-structured interview study', *British Medical Journal* 320, 27 May: 1437–40

Duncan, J. S. (ed) (1981) *Housing and Identity. Cross-cultural Perspectives*, London: Croom Helm Ltd

Dupuis, A. and Thorns, D. C. (1998) 'Home, home ownership and the search for ontological security', *Sociological Review*, 46, 1: 24–47

Dvorak, E. M. and Waymack, M. H. (1991) 'Is it ethical to recruit foreign nurses?', *Nursing Outlook* May/June 120–123

Dyck, I. (1990) 'Space, time and renegotiating motherhood: An exploration of the domestic workplace', *Environment and Planning D*, 8: 459–83

Edgell, S. (1980) *Middle Class Couples*, London: Allen and Unwin

England, K. V. L. (1994) 'Getting personal: Reflexivity, positionality and feminist research', *Professional Geographer*, 46, 1: 80–89.

England, K. V. L. (ed.) (1996) *Who will Mind the Baby? Geographies of Child Care and Working Mothers*, London: Routledge

Epstein, C. F., Seron, C., Oglensky, B. and Saute, R. (1999) *The Part-time Paradox: Time Norms, Professional Life, Family and Gender*, New York: Routledge

Erikson, R. and Goldthorpe J. H. (1992) *The Constant Flux: a Study of Class Mobility in Industrial Societies*, Oxford: Clarendon Press

Etzioni, A. (1969) *The Semi-professions and their Organisations: Teachers, Nurses and Social Workers*, New York: Free Press

European Commission (1998) *Dual Earner Families*, Quarterly Bulletin of the EU Network Family and Work. European Commission Series: Employment and Social AffairsG 44764 Bau: Erkrath

Evenden, L. J. and Walker, G. E. (1993) 'From periphery to centre – the changing geography of the suburbs', in L. G. Evenden and G. E. Walker *The Changing Geography of Canadian Cities*, Montreal and Kingston: McGill-Queen's University Press

Evetts, J. (2000) 'Analysing change in women's careers: Culture, structure and action dimensions', *Gender, Work and Organizations*, 7, 1: 57–67

Felstead, A. and Jewson, N. (2000) *In Work, At Home: Towards an Understanding of Homeworking*, London: Routledge.

Ferber, M. A. and Nelson, J. A. (1993) *Beyond Economic Man: Feminist Theory and Economics*, London: University of Chicago Press

Finch, J. (1987) 'Family obligations and the life course', in A. Bryman, (ed.) *Rethinking the Life Cycle*, London: Macmillan

Findlay, A. M. (1996) 'Skilled transients: The invisible phenomenon. ', in R. Cohen (ed.) *Cambridge Survey of World Migration*, Cambridge: Cambridge University Press 515–22,

Findlay, A. and Gould, W. T. S. (1989) 'Skilled international migration: A research agenda', *Area* 21: 3–11

Findlay, A. M., Li, F. L. N., Jowett, A. J., Brown, M. and Skeldon, R. (1994) 'Doctors diagnose their destination: An analysis of the length of employment of Hong Kong doctors', *Environment and Planning A*, 26: 1605–24

Fishman, R. (1987) *Bourgeois Utopias. The Rise and Fall of Suburbia*, New York: Basic Books

Fitzpatrick, T. (1997) 'A tale of tall cities', *The Guardian*, 6 February.

Fletcher, E. W. L. (1997) 'Home medical students account for less than half of the full registrants Britain requires', *British Medical Journal*, 314: 1278.

Florkowski, G. W. and Fogel, D. S. (1999) 'Expatriate adjustment and commitment: The role of host-unit treatment', *International Journal of Human Resource Management*, 10, 5: 783–807

FOCUS (1996) *Women on the Move*, Conference, March 8–10, Brussels

Forrest, R. and Murie, A. (1987) 'The affluent home owner – labour market position and the shaping of housing history', *Sociological Review* 35, 2: 370–403

Forster, N. (1994) 'The forgotten employees? The experiences of expatriate staff returning to the UK', *International Journal of Human Resource Management*, 5: 405–23

Foucault, M. (1979) *Discipline and Punish: The Birth of the Prison*, Harmondsworth: Penguin

Freely, M. (1999) 'Nice work if you can get it', *Observer*, 4 July 1999, Review Section, 1–2

Futures (2001) *Another country. France versus the Anglo-Saxon economists*, Online. Available HTTP http://www.indsoc.co.uk/futures (accessed 11 June 2001)

Garreau, J. (1991) *Edge City: Life on the New Frontier*, New York: Doubleday

Germain, A. and Rose, D. (2000) *Montrèal: The Quest for a Metropolis*, London: Wiley

Gershuny, J., Godwin, M. and Jones, S. (1994a) 'The domestic labour revolution: A process of lagged adaptation', in M. Anderson, F. Bechhofer and J. Gershuny, (eds) *The Social and Political Economy of the Household*, Oxford: Oxford University Press 151–98

Gershuny, J., Rose, D., Scott, J. and Buck, N. (1994) 'Introducing household panels', in *Changing Households. The BHPS 1990–1992*, Essex: ESRC Research Centre on Micro-Social Change University of Essex

Giddens, A. (1991) *Modernity and Self Identity*, Cambridge: Polity Press

Giddens, A. (1998) *The Third Way: The Renewal of Social Democracy*, Cambridge: Polity

Gillespie, A. (1999) 'The changing employment geography of Britain', in M. Breheny (ed.) *The People: Where will they Work?* London: Town and Country Planning Association 9–28

Gillespie, A., Marvin, S. and Green, N. (2001) 'Bricks versus clicks: Planning for the digital economy', in J. Wilsdon, (ed.) *Digital Futures: Living in a Dot-com World*, London: Earthscan 200–218.

Gilroy, S. (1999) 'Intra-household power relations and their impact on women's leisure', in L. McKie, S. Bowlby and S. Gregory (eds) *Gender, Power and the Household*, Basingstoke: Macmillan 155–72

Ginn, J. and Arber, S. (1995) 'Moving the goalposts: The impact on British women of raising their state pension age to 65', in J. Baldock and M. May (eds) *Social Policy Review* 7, London: Social Policy Association, chapter 11

Ginn, J. and Sandell, J. (1997) 'Balancing home and employment: Stress reported by social services staff', *Work, Employment and Society*, 11, 3: 413–34

Glass, R. (1964) *London: Aspects of Change*, London: Centre for Urban Studies and MacGibbon and Kee

Glazer, P. and Slater, M. (1987) *Unequal Colleagues: The Entrance of Women Into the Professions, 1890–1940*, New Brunswick, Rutgers University Press

Glucksmann, M. (1995) 'Why 'Work'? Gender and the 'Total Social Organisation of Labour', *Gender Work and Organisation* 2, 2: 63–75

Goddard, A. (1999) 'Medical schools win go-ahead', *The Times Higher Education Supplement*, 15 June: 3.

Goldthorpe, J. (1995) 'The service class revisited', in T. Butler and M. Savage (eds) *Social change and the Middle Classes*, London: UCL Press 313–29

Goode, W. (1960) 'A theory of role strain', *American Sociological Review*, 25: 483–96

Gordon, G. E. (1998) 'Work/life – office/home – day/night: Compensating the work-place 'slashes' of the future', *American Compensation Journal*, Special Issue *Work/Life* (www.gilgordon.com/telecommuting/download.htm; accessed 11 October 2000)

Graham, S. and Marvin, S. (1996) *Telecommunications and the City: Electronic Spaces, Urban Places*, London: Routledge

Grandea, N. (1997) 'Migrant workers: A case of Filipina domestic workers in Canada', in H. Gibb (ed.) *Canadian Perspectives of Labor Mobility in APEC*, Vancouver: North-South Institute and Asia Pacific Foundation of Canada

Grant, H. and Oertel, R. (1997) 'The supply and migration of Canadian physicians, 1970–1995: Why we should learn to love an immigrant doctor', *Canadian Journal of Regional Science*, XX: 157–68

Green, A. E. (1997) 'A question of compromise? Case study evidence on the location and mobility strategies of dual career households', *Regional Studies*, 31: 643–59

Green, A. E. Hogarth, T, and Shackleton, R. E. (1999) *Long Distance Living: Dual Location Households*, Bristol: Policy Press.

Gregson, N. and Lowe, M. (1994) *Servicing the Middle Classes: Class, Gender and Waged Domestic Labour in Contemporary Britain*, Routledge: London

Gregson, N. and Lowe, M. (1994a) 'Home-making and the spatiality of daily social reproduction in contemporary middle-class Britain', *Transactions of the Institute of British Geographers*, NS 20: 224–35

Gregson, N. and Lowe, M. (1995) "Too much work?' Class, gender and the reconstitution of middle class domestic labor', in T. Butler and M. Savage (ed.) *Social Change and the Middle Classes*, London: UCL Press 148–68

Grewal, I. (1996) *Home and Harem: Nation, Gender, Empire and the Cultures of Travel*, London: Continuum

Grey, C. (1994) 'Career as a project of the self and labour process discipline', *Sociology*, 28, 2: 479–97

Gross, H. (1980) 'Dual career couples who live apart: Two types', *Journal of Marriage and Family*, August: 567–76

Hakim, C. (2000) *Work-lifestyle Choices in the 21st Century*, Oxford: Oxford University Press

Halfacree, K. H. and Boyle, P. J. (1993) 'The challenge facing migration research: The case for a biographical approach'. *Progress in Human Geography*, 17: 333–48.

Halford, S., Savage, M. and Witz, A. (1997) *Gender, Careers and Organisations: Current developments in Banking, Nursing and Local Government*, Macmillan: Basingstoke

Hamnett, C. (1998) *Winners and Losers: Home Ownership in Modern Britain*, London: UCL Press

Hamnett, C. (1999) *Winners and Losers. Home Ownership in Modern Britain*. London: UCL Press Ltd

Hamnett, C., Harmer, M. and Williams, P. (1991) *Safe as Houses. Housing Inheritance in Britain*, London: Paul Chapman Publishing Ltd

Hannerz, U. (1996) *Transnational connections: Culture, People, Places*, London: Routledge

Hanson, S. and Pratt, G. (1995) *Gender, Work and Space*, London: Routledge

Hardill, I. (1998) 'Trading places: Case studies of the labour market experience of women in rural in-migrant households', *Local Economy*, 13, 2: 102–13.

Hardill, I. (1998a) 'Gender perspectives on British expatriate households', *Geoforum*, 29, 3: 257–68.

Hardill, I. and Green, A. (2001) *Le Manque Démarcation entre le Travail et la Maison: La Dynamique Familiale et les Trajectories chez les Foyers avec Double-carriére*, The Ministère de l' Equipment, des Transports et du Logement, Lettre de commande: F 00–62

Hardill, I. and MacDonald, S. (1998) 'Choosing to relocate: An examination of the impact of expatriate work on dual career households', *Women's Studies International Forum*, 21: 21–9

Hardill, I. and MacDonald, S. (2000) 'Skilled international migration: The experience of nurses in the UK', *Regional Studies*, 34, 7: 681–692

Hardill, I. and Raghuram, P. (1998) 'Diasporic connections: Case studies of Asian women in business', *Area*, 30, 3: 255–61

Hardill, I. and Watson, R. (2001) 'The impact of child rearing upon male and female earnings in dual career households', Paper presented to the Department of Accounting and Finance, University of Glasgow, 14 February 2001

Hardill, I. and Watson, R. (2002, in press) 'An empirical analysis of male and female earnings in dual career households located in the East Midlands region of the UK', *Journal of Industrial Relations*

Hardill, I., Graham, D. T. and Kofman, E. (2001) *Human Geography of the UK: an Introduction*, London: Routledge

Hardill, I., Green, A. E. and Dudleston, A. C. (1997) 'The blurring of boundaries between work and home: Perspectives from case studies in the East Midlands', *Area*, 29, 3: 335–43

Hardill, I., Green, A. E., Dudleston, A. C. and Owen, D. W. (1997a) 'Who decides what? Decision making in dual career households', *Work, Employment and Society*, 11, 2: 313–26

Hardill, I., Dudleston, A. C., Green, A. E. and Owen, D. W. (1999) 'Decision making in dual career households', in L. McKie, S. Bowlby and S. Gregory *Gender, Power and the Household*, Basingstoke: Macmillan 192–206

Hardill, I., MacDonald, S. and Ince, O. (1999) 'Skilled international migration: the importance of EU-qualified nurses in the European Union migratory space', Paper presented at the Regional Studies Association International Conference,

Regional Potentials in an Integrating Europe, University of the Basque Country, Bilbao, Spain 18–21 September 1999

Harding, L. (1997) 'Saudi nurses: compound life: Twisted values in a hidden world', *Guardian*, 25 September 1995: 5

Harkness, (1999) 'Working 9–5?', in P. Gregg and J. Wadsworth (eds) *The State of Working Britain*, Manchester: Manchester University Press 90–108

Harris, R. and Pratt, G. (1993) 'The meaning of home, homeownership, and public policy', in L. Bourne and D. Ley (eds) *The Changing Social Geography of Canadian Cities*, Montreal: McGill-Queen's University Press, 28–97

Harrison, T. (1996) 'Class, citizenship and global migration: The case of the Canadian business immigration program, 1978–1992', *Canadian Public Policy*, XXIl, 1: 7–23

Heitlinger, A. (1999) *Émigré Feminism: Transnational Experiences*, Toronto: University of Toronto Press

Henry, N. and Massey, D. (1995) 'Competitive time-space in high technology', *Geoforum* 26: 49–64

Herman, J. and Gyllstrom, K. (1977) 'Working men and women: inter and intra role conflict', *Psychology of Women Quarterly*, 1: 319–33

Hetherington, P. (1999) 'Suburbia has entered 'crisis of neglect', *Guardian*, 5 February

Hezekiah, J. (1988) 'Colonial heritage and nursing leadership in Trinidad and Tobago', *Journal of Nursing Schools*, 20: 155–8

Hiebert, D. (1999) 'Local geographies of labor market segmentation: Montreal, Toronto and Vancouver', *Economic Geography*, 75: 339–69

Hills, J. (1995) *Inquiry into Income and Wealth Volume 2*, York: Joseph Rowntree Foundation

Hirst, P. and Thompson, G. (1995) 'Globalisation and the future of the nation state', *Economy and Society*, 24: 408–42

HMSO (1994) *Wives Guide to the Army: Army code 61391*, Winchester: HMSO

Hochschild, A. (1990) *The Second Shift*, London: Piatkus

Hochschild, A. (2000) 'Global care chains and emotional surplus value', in W. Hutton and A. Giddens (eds) *On the Edge Living with Global Capitalism*, London: Jonathon Cape

Hogarth, T., Hasluck, C. and Pierre, G. with Winterbotham, M. and Vivian, D. (2000) *Work-Life Balance 2000: Baseline study of Work-Life Balance Practices in Great Britain*, London: Department for Education and Skills.

Holdsworth, D. W. and Simon, J (1993) 'Housing form and use of domestic space' in J. R. Miron (ed.) *House, Home and Community: Progress in Housing Canadians, 1945–1986*, Montreal and Kingston: McGill-Queen's University Press, 188–202

Horrell, S. and Humphries, J. (1995) 'Women's labour force participation and the transition to the male breadwinner family – 1790–1865', *Economic History Review*, XLVIII, 1: 89–117

Huberman, M. (1993) *The Lives of Teachers*, London: Cassell

Hulchanski, J. D. (1993) 'New forms of owning and renting', in J. R. Miron (ed.) *House, Home and Community: Progress in Housing Canadians, 1945–1986*, Montreal and Kingston: McGill-Queen's University Press 64–75

Hunt, J. W. (1982) 'Developing middle managers in a shrinking organization', *Journal of Management Development*, 1, 2: 10–22

IES (2001) *EMERGENCE: Estimation and Mapping of Employment Relocation in a Global Economy in the New Communications Environment*, Brighton: IES.

Ip, D. and Lever-Tracey, C. (1999) 'Asian women in business in Australia', in G. A. Kelson and D. L. DeLaet (eds) *Gender and Immigration*, Basingstoke: Macmillan 59–81

Iredale, R. (1997) *Skills Transfer: International Migration and Accreditation Issues*, Wollongong: University of Wollongong Press

ISIS (2000) *Schools invest in the future as numbers rise again*, Online. Available HTTP: http: //www. iscis. uk. net/ (accessed 26 September 2001)

Jacobsen, J. P. (1994) *The Economics of Gender*, Oxford: Basil Blackwell

Jarvis, H. (1997) 'Housing, labour markets and household structure: Questioning the role of secondary data analysis in sustaining the polarisation debate', *Regional Studies*, 31: 521–32.

Jarvis, H. (1999) 'Situating gender and power relations in the household-locale nexus', Paper presented at WGSG Session *'Is the Future Female?'* the RGS-IBG Annual Conference, University of Leicester, January 1999

Jarvis, H., Pratt, A. C. and Cheng-Chong W. U. P. (2001) *The Secret Lives of Cities: the Social Production of Everyday Life*, Harlow: Prentice Hall

Jinks, C., Ong, B. and Paton, C. (1998) 'Workforce planning: Catching the drift', *Health Service Journal*, 17 September, 108, 5622: 24–6

Johnson, L. C. (1999) 'Bringing work home: Developing a model residentially-based telework facility', *Canadian Journal of Urban Research*, 8, 2: 119–42

Johnson, M. and Bowman, C. C. (1997) 'Occupational prestige for registered nurses in the Asia-Pacific region: Status consensus', *International Journal of Nursing Studies*, 34: 201–07

Johnston, J. H., Salt, J. and Wood, P. (1974) 'Housing and the migration of labour in England and Wales', in M. Savage (1988) 'The missing link? The relationship between spatial mobility and social mobility', *British Journal of Sociology* 39: 554–77

Kanjanapan, W. (1995) 'The immigration of Asian professionals to the United States: 1988–1990', *International Migration Review*, 7–32.

Kloppenburg, J. T. (1989) *Uncertain Victory: Social Democracy and Progressivism in European and American Thought, 1870–1920*, Oxford: Oxford University Press

Knowsley, J. (1999) 'High-fliers quit the city rat race for a stress-free career in garden design', *Sunday Telegraph*, 4 July: 19

Kofman, E. (2000) 'The absence of women and gender relations in studies of European skilled migration', *International Journal of Population Geography*, 6: 45–59

Lakhani, H. (1994) 'The socioeconomic benefits to military families of home-basing of armed forces', *Armed Forces and Society*, 21, 1: 113–28

Layte, R. (1998) 'Gendered equity? Comparing explanations of women's satisfaction with the domestic division of labor', *Work, Employment and Society*, 12, 3: 511–32

Le Snesup, (2001) *Bulletin du Hebomadaire du Syndicat National de l'enseignement Supérieur-FSU*, No 444 du 9 Novembre 2001 SIPE: Paris

Lewis, S. and Cooper, C. (1983) 'The stress of combining occupational and parental roles: A review of the literature', *Bulletin of the British Psychological Society*, 36: 341–5

Lewis, S. and Cooper, C. (1988) 'Stress in dual earner families', *Women and Work*, 3: 139–68

Ley, D. (1993) 'The new middle class in Canadian central cities', in J. Caulfield and L. Peake (eds) *City Lives and City Forms: Critical Research and Canadian Urbanism*. Toronto: University of Toronto Press 1–15

Ley, D. (1993a) 'Past élites and present gentry: Neighbourhoods of privilege in the inner city', in L. Bourne and D. Ley (eds) *The Changing Social Geography of Canadian Cities*, Montreal: McGill-Queen's University Press 214–33

Ley, D. (1996) *The New Middle Class and the Remaking of the Central City*, Oxford: Oxford University Press

Ley, D. (1996a) 'The new middle class in Canadian cities', in J. Caulfield and L. Peake, *City Lives and City Forms: Critical Research and Canadian Urbanism*, Toronto: University of Toronto Press 15–32

Ley, D. F. and Bourne, L. S. (1993) 'Introduction: the social context and diversity of urban Canada', in L. Bourne and D. Ley (eds) *The Changing Social Geography of Canadian Cities*, Montreal and Kingston: McGill-Queen's University Press 3–32

Li, F. L. N., Jowett, A. J., Findlay, A. M. and Skeldon, R. (1995) 'Discourse on migration and ethnic identity: Interviews with professionals in Hong Kong', *Transactions of the Institute of British Geographers*, 20: 342–3–56

Ludingsen, C. (1993) 'Do the continental', *Nursing Times*, 89, 38: 22

McCubbin, H. I. and Lavee, Y. (1986) 'Strengthening army families. A family life cycle stage perspective', *Evaluation and Program Planning*, 9: 221–231

McDonald, J. F. and McMillan, D. P. (2000) 'Employment subcentres and subsequent real estate development in suburban Chicago', *Journal of Urban Economics*, 48: 135–157

McDowell, L. (1991) 'Life without father and Ford: the new gender order of post-Fordism', *Transactions of the Institute of British Geographers NS*, 16: 400–19

McDowell, L. (1997) *Capital Culture*, Oxford: Blackwell

McDowell, L. (1998) *Gender, Identity and Place: Understanding Feminist Geographies*, Oxford: Polity Press

McDowell, L. and Court, G. (1994) 'Gender divisions of labour in the post-Fordist economy: the maintenance of occupational sex segregation in the financial services sector', *Environment and Planning A*, 26: 1397–1418.

McKie, L., Bowlby, S. and Gregory, S. (1999) 'Connection gender, power and the household', in L. McKie, S. Bowlby and S. Gregory, *Gender, Power and the Household*, Basingstoke: Macmillan 3–21

McLaughlin, J. (1997) 'Feminist relations with postmodernism: Reflections on the positive aspects of involvement', *Journal of Gender Studies*, 6, 1: 5–15

Malecki, E. and Bradbury, S. (1992) 'R & D facilities and professional labour: Labour force dynamics in high technology', *Regional Studies*, 26: 123–36

Manrique, C. G. and Manrique, G. G. (1999) 'Third world immigrant women in American higher education', in G. A. Kelson and D. L. DeLaet (eds) *Gender and Immigration*, Basingstoke: Macmillan 103–26

Marchant, K. H. and Medway, F. J. (1987) 'Adjustment and achievement associated with mobility in military families', *Psychology in Schools*, 24: 289–94

Masluch, C. (1987) 'Burnout research in social services: a critique', *Journal of Social Services Research*, 10, 1: 95–105

Massey, D. (1985) *Spatial Divisions of Labour: Social Structures and the Geography of Production*, London: Macmillan

Massey, D. (1994) *Space, Place and Gender*, Cambridge: Polity Press

Mattingly, D. J. (1999) 'Job search, social networks, and local labor market dynamics: the case of paid household work in San Diego, California', *Urban Geography*, 20, 1: 46–74

Mendenhall, M. and Oddou, G. (1985) 'The dimensions of expatriate acculturation: A review', *Academy of Management Review*, 10, 1: 439–87

Micklethwaite, A. and Wooldridge, J. (2000) *A Future Perfect: the Challenge and Hidden Promise of Globalisation*, London: William Heinemann

Mincer, J. (1978) 'Family migration decisions', *Journal of Political Economy* 86: 749–73

MIND (1992) *The Mind Survey: Stress at Work*, London: National Association for Mental Health

Ministry of Defence (1998) *Strategic Defence Review White Paper*, Online. Available HTTP: http://www.mod.uk/index.php3?page=156 (accessed 12 September 2000)

Ministry of Defence (1999) 'People in defence', in *The Strategic Defence Review White Paper 1999* Online. Available HTTP: http://www.mod.uk/index.php3?page=115 (accessed 13 July 2000)

Mitchell, K. (1993) 'Multiculturalism, or the United Colors of Capitalism', *Antipode*, 25: 263–94

Mitchell, K. (1995) 'Flexible circulation in the pacific rim: capitalism in cultural context', *Economic Goegraphy*, 17, 4: 364–382

Mitchell, K. (1997) ' Transnational subjects: constituting the cultural citizen in the era of Pacific Rim capital', in A. Ong and D. Nonini (eds) *Ungrounded Empires: the Cultural Politics of Modern Chinese Transnationalism*, New York: Routledge 228–56

Momsen, J. H. (1984) 'Urbanisation of the countryside in Alberta', in M. F. Bunce (ed.) *The Pressures of Change in Rural Canada*, Geographical Monographs 14, Department of Geography, Atkinson College, York University 160–80

Momsen, J. H. (1999) *Gender, Migration and Domestic Service*, London: Routledge

Monk, J. (1983) 'Asian professionals as immigrants: The Indians in Sydney', *Journal of Cultural Geography*, 4: 1–17

Moore, H. (1988) *Feminism and Anthropology*, Minneapolis: University of Minnesota Press

Morgan, D. (1999) 'Gendering the household: Some theoretical considerations', in L. McKie, S. Bowlby and S. Gregory (eds) *Gender, Power and the Household*, Basingstoke: Macmillan 22–40

Morgan, P. (1989) 'People', *Accountancy*, January: 94–5

Morris, L. (1990) *The Workings of the Household*, Oxford: Polity Press

Munro, M. and Madigan, R. (1998) 'Housing strategies in an uncertain market', *The Sociological Review*, 714–734

Myrah, M. (1997) 'High skill workers: labor mobility in the high tech sector', in H. Gibb (ed.) *Canadian Perspectives of Labor Mobility in APEC*, Vancouver: North-South Institute and Asia Pacific Foundation of Canada

Newell, S. (1993) 'The superwoman syndrome: Gender differences in attitudes towards equal opportunities at work and towards domestic responsibilities at home', *Work, Employment and Society*, 7, 2: 275–289

Norton, C. (1999) 'Workaholic Britain is turning into the 'grab and go' society', *Independent*, 27 July: 6

Oerton, S. (1997) "Queer housewives': Some problems in theorising the division of domestic labor in lesbian and gay households', *Women's Studies International Forum*, 20, 3: 421–30

O'Hara, B. (2001) 'A prescription for Canada's overtime addiction', *Vancouver Sun, Province and Nation section*, A15

Ong, A. (1999) *Flexible Citizenship: the Cultural Logics of Transnationality*, Durham and London: Duke University

Ong, P. M., Cheng, L. and Evans, L. (1992) ' Migration of highly educated Asians and global dynamics', *Asian and Pacific Migration Journal*, 1: 543–67.

ONS and Equal Opportunites Commission (1998) *Social Focus on Women and Men*, London: HMSO

Oommen, T. K. (1989) 'India: brain drain or the migration of talent', *International Migration*, XXVII: 411–2

Pahl, J. (1989) *Money and Marriage*, London: Macmillan

Parekh, B. (1994) 'Some reflections on the Hindu diaspora', *New Community*, 20, 4: 603–20.

Peck, J. A. (1996) *Work Place: the Social Regulation of Labor Markets*, New York: Guilford Press

Pe-Pua, R., Mitchell, C., Iredale, R. and Castles, S. (1996) *Astronaut Families and Parachute Children: the Cycle of Migration between Hong Kong and Australia*, Canberra: Australian Government Publishing Service

Pesman, R. (1996) *Duty Free: Australian Women Abroad*, Oxford: Oxford University Press

Philliber, W. W. and Vannoy-Hillier, D. (1990) 'The effects of husband's occupational attainment on wife's achievement', *Journal of Marriage and Family*, 52: 323–29

Phillips, M. (1998) 'The restructuring of social imaginations in rural geography', *Journal of Rural Studies*, 14, 2: 121–65.

Phillips, S. (1982) 'Career exploration in adulthood', *Journal of Vocational Behavior*, 20, 129–40

Phizacklea, A. and Ram, M. (1996) 'Being your own boss: Ethnic minority entrepreneurs in comparative perspective', *Work, Employment and Society*, 10: 319–39

Poole, M. and Isaacs, D. (1997) 'Caring: A gendered concept', *Women's Studies International Forum*, 20, 4: 529–36

Pratt, G. (1981) 'The house as an expression of social worlds' in J. S. Duncan (ed.) *Housing and Identity. Cross-cultural Perspectives*, London: Croom Helm 138–80

Rabinovitz, J. (1999) 'Long distance lifestyle: big-time commute offers some big-time benefits', *Sacramento Bee* 7 March, E1.

Rao, K. (2000) Technology, *The Banker*, 890: 102–7

Rapoport, R. and Rapoport, R. N. (1976) *Dual-Career Families Re-examined*, London: Martin Robertson

Redman, T., Wilkinson, A. and Snape, E (1997) 'Stuck in the middle? Managers in building societies', *Work, Employment and Society*, 11, 1: 101–14

Reeves, R. (2001) *Happy Monday: Putting the Pleasure back into Work*, Harlow: Pearson Education

Reskin, B. and Padavic, I. (1994) *Women and Men at Work*, California: Pine Forge Press

Reynolds, N. (2000) 'Opera star cancels to care for his baby', *Daily Telegraph* 7 October: 1

Riska, E. and Wegar, K. (1993) *Gender, Work and Medicine*, London: Sage

Robinson, S. (1995) 'Florence of Arabia', *Nursing Times* 91, 42: 46–47

Rose, D. (1996) 'Economic restructuring and the diversification of gentrification in the 1980s. A view from a marginal metropolis', in J. Caulfield and L. Peake (eds) *City Lives and City Forms: Critical Research and Canadian Urbanism*, Toronto: University of Toronto Press 131–72

Rose, D. and Carrasco, P. (2000) 'Gender divisions of unpaid household labour in Central Canadian cities in 1996: The case of dual earner couples', Paper presented at the Annual meeting of the *Canadian Regional Science Association*, Toronto 3–4 June 2000

Rose, N. (1989) *Governing the Soul: the Shaping of the Private Self*, London: Routledge

Rossi, P. H. (ed.) (1980) *Why Families Move*, London: Sage (2nd edition)

Ryten, E., Thurber, A. D. and Buske, L. (1998) 'The class of 1989 and physician supply in Canada', *Canadian Medical Journal*, 158: 723–8

Sacramento Bee (The) (1999) 'Long distance lifestyle: big-time commute offers some big-time benefits', 7 March, E1.

Safilios-Rothschild, C. (ed.) (1972) *Towards a Sociology of Women*, Lexington, Mass: Xerox

Said, E. (1994) *Culture and Imperialism*, London: Vintage

Salt, J. (1988) 'Highly skilled migrants, careers and international labour markets', *Geoforum*, 19, 387–399

Saunders, P. (1989) 'The meaning of home in contemporary English culture', *Housing Studies*, 4, 3: 177–192

Savage, M. (1988) 'The missing link? The relationship between spatial mobility and social mobility', *British Journal of Sociology*, 39: 554–77

Savage, M. (1995) 'Class analysis and social research', in T. Butler and M. Savage (eds) *Social Change and the Middle Classes*, London: UCL Press 15–25

Sawada, A. (1997) 'The nurse shortage problem in Japan', *Nursing Ethics*, 4, 3: 245–52

Scase, R., Scales, J. and Smith, C. (1998) *Work Now Pay Later: the Impact of Working Long Hours on Health and Family Life*, Working Paper of the ESRC Research Centre on Micro-Social Change, University of Essex.

Schoenberger, E. (1997) *The Cultural Crisis of the Firm*, Oxford: Blackwell

Scott, P. (1966) *The Jewel in the Crown*, London: Heinemann

Scullion, H. (1994) 'Staffing policies and strategic control in British multinationals', *International Studies of Management and Organization*, 24, 3: 86–104

Seavers, J. (1999) 'Residential relocation of couples: The joint decision-making process considered', in P. Boyle and K. Halfacree (eds) *Migration and Gender in the Developed World*, London: Routledge 151–71

Seccombe, I., Buchan, J. and Ball, J. (1993) 'Nurse mobility in Europe: Implications for the United Kingdom', *International Migration*, XXXI, 1: 125–48

Sennett, R. (1998) *The Corrosion of Character*, New York: Norton

Sieber, S. (1974) 'Toward a theory of role accumulation', *American Sociological Review*, 39: 567–78

Sizoo, E. (1997) *Women's Lifeworlds: Women's Narratives on Shaping their Realities*, London: Routledge

Skaburskis, A. and Fullerton, C. (1998) 'Research note: The lure of Ottawa's rural region', *Canadian Journal of Urban Research*, 7, 2: 244–59

Smart, J. (1997) 'Borrowed men on borrowed time; Globalisation, labor migration and local economies', in Alberta *Canadian Journal of Regional Science*, XX: 141–56

Smart, A. and Smart, J. (1993) 'Monster homes: Hong Kong immigration to Canada, urban conflicts, and contested representations of space', in J. Caulfield and L. Peake (eds) *City Lives and City Forms: Critical Research and Canadian Urbanism*, Toronto: University of Toronto Press 33–46

Smith, C. R. (1996) 'Dual careers, dual loyalties: Management implications of the work/home interface', *Asia Pacific Journal of Human Resources*, 34: 19–29

Stacey, J. (1998) *Brave New Families: Stories of Domestic Upheaval in Late Twentieth Century America*, New York: Basic Books

Staeheli, L. A. and Lawson, V. A. (1994) 'A discussion of women in the field: The politics of feminist fieldwork', *Professional Geographer*, 46, 1: 96–102

Stanworth, C. (1998) 'Telework and the information age', *New Technology, Work and Employment* 13, 1: 51–62

Stanworth, C. (2000) 'Women and work in the information age', *Gender, Work and Organization* 7, 1: 20–32

Steele, M. (1993) 'Income, prices and tenure choice', in J. R. Miron (ed.) *House, Home and Community: Progress in Housing Canadians, 1956–86*, Montreal and Kingston: McGill-Queen's University Press 41–63

Sullivan, W. (1995) *Work and Integrity: the Crisis and Promise of Professionalism in America*, New York: Harper Collins

Super, D. (1957) *The Psychology of Careers*, New York: Harper and Row

Susser, M. W. and Watson, W. (1975) *Sociology in Medicine*, London: Oxford University Press (3rd edition)

Sutherland, V. and Cooper, C. (1990) *Understanding Stress: a Psychological Perspective for Health Professionals*, Suffolk: Chapman and Hall

Suutari, V. and Brewster, C. (1999) 'International assignments across European boundaries', in C. Brewster and H. Harris (eds) *International HRM*, London: Routledge

Swain, H. (1999) 'UK benefits from brain drain', *The Times Higher Education Supplement*, 16 July: 60

Telecottage Association (TCA) (1999) *Virtual call centre research*, Telenews Online. Available HTTP: http://www.tca.org.uk/news.htm (accessed 29 Oct, 1999)

Thornley, C. (1996) 'Segmentation and inequality in the nursing workforce', in R. Crompton, D. Gallie and K. Purcell (eds) *Changing Forms of Employment: Organisations, Skills and Gender*, London: Routledge 160–81

Thrift, N. and Leyshon, A. (1991) 'In the wake of money', in L. Budd and S. Whimpster (eds) *Global Finance and Urban Living*, London: Routledge

Torbiorn, I. (1982) *Living Abroad: Personal Adjustments and Personnel Policy in Overseas Settings*, New York: John Wiley

Townsend, A. (1997) *Making a Living in Europe*, London: Routledge

Tsay, C. L. (1995) 'Taiwan', *ASEAN Economic Bulletin*, 12, 2: 175–90

Tyner, J. A. (2000) 'Global cities and circuits of global labor: The case of Manila, Philippines', *Professional Geographer*, 52, 1: 61–73

Tysome, T. (1999) 'Demand for degrees outstrips supply', *The Times Higher Education Supplement*, 19 March: 8–9.

Ukcosa (Council for International Education) (2000) in J. Bown 'The give-and-take case for overseas students', *The Times Higher Education Supplement*, Opinion 23 June

United Kingdom Central Council (UKCC) (1991) *Statistical Analysis of the UKCC's Professional Register 1 April 1990 to March 31 1991*, London: United Kingdom Council for Nursing, Midwifery and Health Visiting

United Kingdom Central Council (UKCC) (1998) *Statistical Analysis of the UKCC's Professional Register for 1 April 1997 to March 31 1998*, London: United Kingdom Council for Nursing, Midwifery and Health Visiting

United Kingdom Central Council (UKCC) (2001) *Statistical Analysis of the UKCC's Professional Register for 1 April 2000 to March 31 2001*, London: United Kingdom Council for Nursing, Midwifery and Health Visiting

Utley, A. (1998) 'Women lose out on leisure', *The Times Higher Education Supplement*, 7 August: 5

Valdez, R. and Gutek, B. (1987) 'Family roles: A help or a hindrance for working women?', in B. Gutek and L. Larwood (eds) *Women's Career Development*, Beverly Hills: Sage 157–69

Valentine, G. (1999) 'Doing household research: Interviewing couples together and apart', *Area* 31, 1: 67–74

Van Loon, J. (2002) *Risk and Technological Culture*, London: Routledge

Van Manen, M. (1990) *Researching Lived Experience: Human Science for an Action Sensitive Pedagogy*, New York: State University Press

Verney, R. E. (1957) *The Student – the Philosophy of Sir William Osler*, London: Livingstone

Vertovec, S. (1999) 'Conceiving and researching transnationalism', *Ethnic and Racial Studies*, 22, 2: 447–62

Walby, S. (1989) *Theorizing Patriarchy*, Oxford: Basil Blackwell

Walby, S. and Greenwell, J., with MacKay, L. and Soothill, K. (1994) *Medicine and Nursing: Professions in a Changing Health Service*, London: Sage

Walsh, N. P. (2001) 'Millions are driven to attack their computers', *Observer*, 10 June

Walter, B. (1989) *Irish Women in London: The Ealing Dimension*, London: Ealing Women's Unit

Walters, J. (2001) 'The flexible family? Recent immigration and 'astronaut' households in Vancouver, British Columbia', *University of British Columbia RIM*, Working paper series, 01–02

Walton-Roberts, M. (1998) 'Three readings of the turban: Sikh identity in Greater Vancouver', *Urban Geography*, 19, 4: 61–73

Wharton, C. S. (1994) 'Finding time for the second shift: The impact of flexible working on women's double roles', *Gender and Society*, 8, 2: 180–205

Wheelock, J., Baines, S. and Oughton, E. (2000) *Individual Economic Responsibility or Social Well-being? At the Interface between Economics and Social Policy*, Berlin: EAPE 2000 Conference

Wilcock, A. (1996) 'Flying high in America', *Nursing Times*, 92, 6: 16–17

Wilkinson, H. (1999) 'Honey, I'm (still) home', *Guardian*, 14 July: 6

Wilson, R. (1999) 'The frustrating career of the trailing spouse', *The Chronicle of Higher Education*, 19 March: 12–13.

Windham/National Foreign Trade Council (1995) *Global Relocation Trends 1995 Survey Report*, New York: Windham International

Winfield, F. E. (1985)) *Commuter Marriage: Living Together, Apart*, New York: Columbia University Press

Wolff, J. (1995) *Resident Alien: Feminist Cultural Criticism*, Cambridge: Polity Press

Woodward, T. (1999) 'These young doctors cost £28 million to train. Now 34 are abandoning the NHS. Why?', *Daily Mail*, 31 July: 16–17.

Yeoh, B. S. A. and Khoo, L. M. (1998) 'Home, work and community: Skilled international migration and expatriate women in Singapore', *International Migration*, 36, 2: 159–86

Yeoh, B. and Willis, K. (2000) 'On the 'regional beat': Singapore men, sexual politics and transnational spaces', Presentation given at the Association of American Geographers 96th Annual Meeting, 4–8 April, Pittsburgh, Pennsylvania.

Young, M. and Wilmott, P. (1973) *The Symmetrical Family*, New York: Penguin.

Index